The Moth Snowstorm

Also by Michael McCarthy

Say Goodbye to the Cuckoo

The Moth Snowstorm

Nature and Joy

MICHAEL McCARTHY

JOHN MURRAY

First published in Great Britain in 2015 by John Murray (Publishers)
An Hachette UK Company

1

© Michael McCarthy 2015

The right of Michael McCarthy to be identified as the Author
of the Work has been asserted by him in accordance with the
Copyright, Designs and Patents Act 1988.

A CIP catalogue record for this title is available from the British Library

ISBN 978-1-444-79277-5
Ebook ISBN 978-1-444-79278-2

Typeset in Bembo by Palimpsest Book Production Limited,
Falkirk, Stirlingshire

Printed and bound by Clays Ltd, St Ives plc

John Murray policy is to use papers that are natural, renewable
and recyclable products and made from wood grown in sustainable
forests. The logging and manufacturing processes are expected to
conform to the environmental regulations of the country of origin.

John Murray (Publishers)
Carmelite House
50 Victoria Embankment
London EC4Y 0DZ

www.johnmurray.co.uk

To the memory of Norah

What would the world be, once bereft
Of wet and of wildness? Let them be left,
O let them be left, wildness and wet;
Long live the weeds and the wilderness yet.

Gerard Manley Hopkins, 'Inversnaid'

Contents

I

A Singular Window

In the summer of 1954, when Winston Churchill was dwindling into his dotage as British prime minister, the beaten French were withdrawing from Indochina, and Elvis Presley was beginning to sing, my mother's mind fell apart. I was seven and my brother John was eight. Norah our mother was forty years old and was a teacher, although she had given up teaching when we were born. She came from a poor family but had won scholarships and been well educated, and was well read with a literary bent; she had corresponded with Pádraic Colum, he who wrote *She Moved Through the Fair*, after the Irish poet had struck up a friendship with her father, a merchant marine steward, on a transatlantic voyage; she had a novel sketched out (it was about a paediatrician who was wonderful with children and terrible with everybody else). She was gentle and kind to a fault, entirely unselfish and wholly honest, and a deeply religious Catholic.

Her mind had begun to fray during the long absences of my father Jack, who was a radio officer on the *Queen Mary*. These were the final glory days of the Cunard ocean liners, and it might be said that his life sailing regularly between Southampton and New York was glamorous; certainly it was more so than the life he found when he came home to our small terraced house in Birkenhead, the town across the River Mersey from Liverpool, for two weeks every three months. He was not loving, either as a husband or a father – he did not know how to love – although he was not a bad man; but he covered up his lifelong insecurity with bluster, which too often turned into bad temper. The ten-year marriage, doctors clinically recorded later, had been 'moderately happy'.

In her long isolations, bringing us up, my mother was closely supported by her well-meaning but bossy sister Mary and Mary's biddable husband Gordon, who were childless, but in 1953 Mary and Gordon spent several months in America with friends – they contemplated emigrating – and it was during this time, when she was even more cut off, that my mother's psyche began to wander. When Mary and Gordon returned they saw a change in her, and as 1953 became 1954 Norah began to behave strangely; she went missing for a day and was found twenty miles away, having walked alone for many hours. She grew increasingly troubled as the year went on and the climax came in the summer when she threw herself upon the mercy of Mother Church, which sixty years ago straitly governed the lives of the Irish Catholic families of Merseyside, and Canon Quinn, iron-fisted ruler of the parish of Our Lady, decreed she should be sent to an asylum.

No one could stand against this because no one understood what was wrong with her, other than that her mind was obviously in turmoil and that she was greatly distressed. As were the rest of the family: Jack and Mary and Gordon and other close relatives were not only at a loss but also ashamed, since this was well before R. D. Laing turned mental illness into martyrdom, and the best my father could do was borrow the money for her to go as a private patient, and off to the mental hospital she went, from whence, as was pointed out to me much later, you did not, in those days, often emerge.

So it might have been with her, for the doctors who treated her were equally perplexed and were unable to offer her anything other than electro-convulsive therapy, high-voltage shocks to the head, before each session of which she thought she would die; but after several weeks, one day, just like that, a particularly perspicacious psychiatrist suddenly grasped it, and made an opening into her anguished spirit that paved the way for her eventual recovery (circumstances I discovered forty-one years later when I contacted the hospital and found that, miraculously, they still held her notes, which they released to me).

Thus, after nearly three months, Norah came back; but when she had gone away, in the August, our family life had been blown to bits. My father could not hold it together; he was a distant, irritated figure, impatient with John and me, and anyway he was still away at sea, so Mary took us in charge. She sold our house – she sold our house! – took us into her home in Bebington, a nearby suburb, and offered us kindness, but the damage was done. John's stability was destroyed. Decades afterwards I met the woman who had been his teacher, Miss Dowling, and she told me that when he came into the class that September he would sit staring ahead and bending wooden rulers between his hands until they broke, one after another – he was eight years old – and she said she told the class, we all have to be kind to John and not make fun of him, because he is very upset because his mum has gone away. From then on his instability was life-long and meant that he struggled to cope with all aspects of existence (apart from the piano keyboard, at which he was in charge). He had paid the price for understanding. Not rationally. Nobody understood rationally what had happened to Norah, no one at all, until her hospital notes reached me half a lifetime later. But John, a sensitive boy that little bit older, had understood it fully with his emotions; he felt what she was going through and defended her fiercely against the wallowing, helpless, uncomprehending adults, and the knowledge was more than he could bear.

3

I was worse than uncomprehending; I was indifferent. At seven years old, I was not in the least bit concerned that I had lost my mother. How bizarre that seems, written down. Many years on, when I began to talk about it, to try to sort it all out, I learned that this was a Coping Strategy. Golly, I thought. Did I have a Coping Strategy? All I remember having is nothing. Being not bothered, not in the slightest, that she had gone away with no promise of return; and this attitude slumbered inside me through childhood, adolescence and long into manhood, until my mother died, my mother with whom I had by now built bridges and come to adore before all others, and I found that I could not mourn her. Just as I had been indifferent when she first went away, I found to my consternation that I was indifferent now when she went away for ever – and the life I had blithely put together on top of the gaping cracks, pretending they were not there, began to unravel, and I set out on the long road to somewhere else.

But in August 1954 there was no difficulty. There was no emotion. John found it difficult in the extreme, he was upset daily, he screamed out loud, but for me there was nothing, it was as if my soul had been ironed flat on a board, with not a ripple or a wrinkle in it, when we took up residence with Mary and Gordon. They lived in a short cul-de-sac called Sunny Bank, and as it was suburban, it was considerably greener than our Birkenhead terrace; the houses had front gardens, and in one of them, two doors away, hanging over the wall, was a buddleia.

There are plants famed for their healing properties and plants notorious as poisons, and others familiar because we consume them and some because we use their fibres, but there are not many specifically known for being powerful attractants of one kind of wildlife. Yet such is *Buddleia davidii*, discovered in the mountains of China in 1869 by the wandering French Jesuit-naturalist Père David (he who introduced to Europe, among much other exotic wildlife including the deer that bears his name, the first giant panda); and I stumbled upon this property of the plant when on a bright morning, soon after we arrived, I ran out of the house into Sunny Bank to play and encountered the tall bush covered in jewels, jewels as big as my seven-year-old hand, jewels flashing dazzling colour combinations: scarlet and black, maroon and yellow, pink and white, orange and turquoise. The buddleia was crawling with butterflies. They were mainly the nymphalid quartet of late summer in England, red admirals, peacocks, small tortoiseshells and painted ladies, the ones which take on fuel in August to hibernate or migrate for the winter – the gaudiest of all the British Lepidoptera, bumping into each other on the plump purple flower spikes in their greedy quest for nectar.

I gazed up at them. I was mesmerised. My eyes caressed their colours like a hand stroking a kitten. How could there be such living gems? And every morning in that hot but fading summer, as my mother suffered silently and my brother cried out, I ran to check on them, never tiring of watching these free-flying spirits with wings as bright as flags which the buddleia seemed miraculously to tame, to keep from visiting other flowers, to enslave on its own blooms by its nectar's unfathomable power. I could smell it myself, honey-sweet, but with the faintest hint of a sour edge. Drawing them in, the wondrous visitants. Wondrous? Electrifying, they were. Filling the space where my feelings should have been. And so, through this singular window, when I was a skinny kid in short pants, butterflies entered my soul.

That we might love the natural world, as opposed to being wary of it, or instinctively conscious of its utility, may be thought of as a commonplace; but over the years it has increasingly seemed to me a remarkable phenomenon. For after all, it is only our background, our context, the milieu from which, like all other creatures, we have emerged. Why should it evoke in us any emotion beyond those, such as fear and hunger, that are needed for survival? Can an otter love its river? And yet it is the case, that the natural world can offer us more than the means to survive, on the one hand, or mortal risks to be avoided, on the other: it can offer us joy.

Although I strongly feel that this is one of the greatest things in our lives – never more important than now – it seems quite mysterious in its origins, and certainly in the force it can exercise. To be able to be swept up, carried away, by an aspect of nature such as butterflies; tell me, is that something in nature itself, or is it something in us? Once, Christianity offered a ready explanation: our joy in the beauty and life of the earth was our joy in the divine work of its creator. But as Christianity fades, the undeniable fact that the natural world can spark love in us becomes more of an enigma.

You can see far more easily why it engenders some other powerful emotions, with, for example, the big beasts. The first big beast I ever saw in the wild was a black rhino, in Namibia. It was about a hundred yards away, a ton of double-horned power glaring straight at me with nothing but low scrub between us; and although I knew it had poor eyesight, it was twitching its ultra-sensitive ears like revolving radar antennae, trying to pick me up and draw a bead on me, and I was transfixed: my heart pounded, my mouth dried, I looked around for shelter. But if I was afraid, there was a stronger and stranger feeling coursing through me. I felt in every way more alive. I felt as alive as I had ever been.

The next day I saw an African buffalo for the first time, a great black mass of menace which made me even more nervous than the rhino had, yet I experienced precisely the same sensation: mixed in with the anxiety, with the fear of being killed,

and buffalos will kill you, was the feeling in the animal's prox-
imity of living more intensely, of somehow living almost at
another level. And when later that day in a dry riverbed I saw,
close to, my first wild elephant, the most dangerous of them
all, I felt again, intermingled with the wariness, something akin
to passion.

They are surely very old, these feelings. They are lodged deep
in our tissues and emerge to surprise us. For we forget our
origins; in our towns and cities, staring into our screens, we
need constantly reminding that we have been operators of
computers for a single generation and workers in neon-lit offices
for three or four, but we were farmers for five hundred generations, and before that hunter-gatherers for perhaps fifty
thousand or more, living with the natural world as part of it as
we evolved, and the legacy cannot be done away with.

It is to those fifty thousand generations that our fascination
with the big beasts harks back; their magnificence triggers an
awe in us, the still surviving awe of our ancestors who pursued
them, full of fear and hope, piously painting their images on
the walls of caves. On the rock faces of Lascaux and Chauvet,
where the fear and hope coalesce into worship, we have astonishing insights into a world of long-gone people whose lives
revolved around dangerous animals and their slaughter, and who
must therefore have lived, with mortality ever present, at that
elevated and passionate level we still sense when we come up
against the great beasts ourselves, in their natural surroundings.

Yet a stray thought plays about my mind, haunts its corners,
refuses to leave: it must also be the case that the hunter-gatherers
saw butterflies. Were they indifferent? All of them? Even to
swallowtails? Somehow I doubt it. I think the point must have
arrived where such unlikely, brilliant beings could not but register
with observers, even those obsessed with survival and violence
and death – that a moment must have come in prehistory when
someone, for the very first time, waited for a swallowtail to

settle, the better to look on it, and then marvelled at what was there in front of them.

Childhood does not conform to a pattern, though we tend to assume it does. We have templates in our minds for human lives, how they should begin, come to maturity, and end; in short, how they should play out; and often we try to make sense of our own experience by aligning it with one template or another, and seeing how far it differs or corresponds. Yet in reality, of course, the forms of our experience are infinite.

I have lived most of my life now. I have been fortunate in learning to repair a fair amount of the damage of the early years, and maybe even more important, in learning to live in peace alongside what could not be repaired: I said to John once, as we came out of another tense Monday morning session of untangling ancient anguishes, the Greeks gave us politics, and history, and the theatre, but they never managed to come up with family therapy, and he smiled and agreed. And this idea of living in peace alongside abnormality is perhaps what has allowed me to accept the strange circumstance, that it was in a time of turmoil, involving great unhappiness, that I first became attached to nature; that while my boyhood bond with my mother was being rent asunder, I was preoccupied with insects.

For I do accept it; I was seven, and not to be blamed, and besides, the allure of butterflies has worked its charm on far more significant minds than mine, although I would say, perhaps not on all that many minds more susceptible at a given moment: having shut out what was really happening, my spirit was an empty tablet, open to impressions, and the scarlet and black of the red admiral painted themselves on it indelibly, as did the brilliant colours of its cousins. I was gripped by a fervent enthusiasm; I babbled to Mary of it, and she obligingly bought me *The Observer's Book of Butterflies* as part of her campaign to be accepted as mother substitute. ('There'll be lots of treats!' she had announced as she scooped us up and carried us off.)

Turning its old-fashioned pages, while my real mother was somewhere else, I didn't know where, I began to marvel even more at pictures of species I could only dream about, some of them not only magnificent in appearance but possessed of awe-inspiring names: The Duke of Burgundy! The purple emperor! The Queen of Spain fritillary! O brave new world, that hath such creatures in it! And over the following weeks and months this enthusiasm flourished and deepened, even though it was a year of trauma, for in October, after the one perceptive doctor had made it possible, Norah left hospital, well enough to leave, but by no means properly well, and came to live with us all in Sunny Bank until we could once more acquire a house of our own.

This was something Mary hadn't bargained for. She had thought her sister gone for good. (Probably most people had.) She was perfectly prepared, perhaps even secretly delighted, to take on two young boys, an instant family of her own; but the new situation, which in effect meant sharing her home with a separate family that was now badly disturbed, was a quite different prospect, albeit one she was obliged to accept. She had sold our house, after all.

For her part, Norah, who had the most sensitive of spirits and was in no way robust, had not only been plunged into purgatory by her original ailment, but had been profoundly shocked by the experience itself of being incarcerated and deprived of her children. (She told me many years later that she thought she would never see us again, and the only solace she found was in the two anguished sonnets of Gerard Manley Hopkins, the ones which begin 'No worst, there is none' and 'I wake and feel the fell of dark, not day', because at least they showed her that others had been in the place where she now found herself, and had survived.) Her equilibrium was shattered, and when she came out she was fearful and suspicious, believing people were talking about her in the street. (Perhaps they were.) Most difficult of all was to return, not to the care of a loving husband – he was off in the North Atlantic somewhere, dining at the captain's table – but to go as a lodger to the house of a sister who, she might well think in her very unsettled state, had tried to steal her children.

The dynamics were explosive. Forced together, Mary and Norah were at each other's throats. It was a year filled with shouting, with tormented quarrels and upset which drove John frantic with distress. I can remember flashes of it. I can remember Norah trying to push Mary on the stairs with a strange look on her face, and Gordon screaming at her *Something something, Sister!* But once again it all washed over me: coping strategy or whatever it was, what concerned me was the Butterfly Farm in Bexley, Kent, of L. Hugh Newman, Esq. – the curious caprice of the initial in front of the name somehow adding to its mystique – which would supply you with caterpillars of the most splendid British butterflies you could wish for, to be nursed into metamorphosis in your own home. In his catalogue, Mr Newman referred to the caterpillars and adult butterflies he sold as 'livestock', something else I found curiously engaging, and as the spring of 1955 came into view I sent off my five-shilling postal order for some livestock and duly received a cardboard tube containing two purple emperor caterpillars on a branch of sallow, their food plant. They died before their metamorphosis could take place. So did a second pair. I was clearly doing something wrong. But it wasn't for lack of ardour. Butterflies had indeed entered my soul. They were beings I felt intensely bound to – I could have described to you, then, the row of tiny turquoise crescents on the lower edges of the wings of the small tortoiseshell – and I suppose I might have gone on to become a lifelong butterfly obsessive, narrowly and compulsively preoccupied to the exclusion of all else, like Frederick Clegg in John Fowles' *The Collector*, had not my mother shown me the way to a wider world.

She did so shortly after she and my father finally managed to find another home of their own, in the November of 1955, and we could leave Sunny Bank after all the discord and try to begin again, and she did so with a book. It was a Christmas present that year, prompted I imagine by my butterfly enthusiasm; but whereas Mary might have found me another book on Lepidoptera, Norah chose something else, and I wonder now what sure instinct led her to this, the first real story I encountered, with fully formed characters and a narrative; for I engaged with it at once.

It was an epic, in the old-fashioned, precise sense of the term: a long account of heroic adventures. But it was not large-scale, in the way that *The Iliad* and *The Odyssey* are large-scale epics, mainly because its heroes were gnomes. It was called *The Little Grey Men*, and its author signed himself merely by initials, 'BB'; his real name was Denys Watkins-Pitchford, although it was years before I found this out.

I was from the first page lost in the world of its principal characters, Dodder, Baldmoney, and Sneezewort (all named after rather uncommon English wild flowers). They were very small people, between a foot and eighteen inches tall, with long flowing beards; Dodder, the oldest, had a wooden leg. But they were different from the sort of gnomes you might expect to come across in the genre of High Fantasy which has so obsessed us in recent years, in *Harry Potter* and *The Lord of the Rings* and their imitators. They had no magical powers. They were grounded not in fantasy but in realism. Although they were able to converse with the wild creatures around them – the author's one concession to the idea of gnomic difference – they lived, and struggled to live, in the world just as we do, concerned about finding enough food and keeping warm. But there was more: they were a dying race. They were the last gnomes left in England.

I remember the shiver I experienced when I first read those words. I think it was an inchoate sense, even in a boy of eight, of the transfixing nature of the end of things. It was clear that they could not survive the creeping urbanisation and the modernisation of agriculture which even then were starting to spread across the countryside. They were anachronisms. The world had moved on from them: like Butch Cassidy and the Sundance Kid, their time was done. So much the braver, then, their decision to undertake a great adventure, to make an expedition to find their long-lost brother Cloudberry – ah, Cloudberry! So sad! – who had never returned after setting out one day to discover the source of the small Warwickshire river, the Folly Brook, on the banks of which they lived, in the capacious roots of an oak tree.

I was wholly captivated by their quest, and by its unexpected denouement; I was likewise captivated by *Down the Bright Stream*, the sequel, which I asked for and was given for Christmas the following year. (In the second book, the gnomes' existential crisis reaches its climax; they address it in a most original way, ultimately successfully.) But I took in more than the story. I internalised, at first reading, the milieu in which the adventures took place. It was the very opposite of the milieu of *The Lord of the Rings*, with its dark lords and wizards, its fortresses and mountains, its vast clashing armies; it was merely Warwickshire, leafy Warwickshire, Shakespeare's county, and the Folly Brook, with its kingfishers and otters and minnows, and its kestrels hovering above, a small and intimate and charming countryside with its small and intimate and charming creatures, vivid in their lives and their interactions; and I fell in love with them, and I fell in love with the natural world. I went beyond butterflies into the fullness of nature.

I was immensely lucky: I discovered it right at the end of what one might call the time of natural abundance (at least, in

my own country of England). It was several years before intensive farming would take a stranglehold on the land – before
the detestable tide of organochlorine and organophosphate pesticides began to wash over it and burn it like acid burns a body
– and something taken for granted but wonderful persisted still:
natural profusion. There had been *lots* of butterflies on the
buddleia, in the August when I first encountered them.

This is not childhood seen through rose-tinted binoculars. I
remember it clearly. It was somehow at the heart of the attraction. I don't think I could have been affected in the same way
by a solitary red admiral, marvel of creation though it is. There
were lots of many things, then. Suburban gardens were thronged
with thrushes. Hares galumphed across every pasture. Mayflies
hatched on springtime rivers in dazzling swarms. And larks filled
the air and poppies filled the fields, and if the butterflies filled
the summer days, the moths filled the summer nights, and sometimes the moths were in such numbers that they would pack
a car's headlight beams like snowflakes in a blizzard, there would
be a veritable snowstorm of moths, and at the end of your
journey you would have to wash your windscreen, you would
have to sponge away the astounding richness of life.

It was to this world, the world of the moth snowstorm, that
I pledged my youthful allegiance.

Yet the twenty-first century will be terrible for the natural
world to which as a young boy I became so bound.

I am a baby boomer; I am of that generation born in the
rich West in the aftermath of the Second World War, the generation which came to adulthood in the explosion of new freedoms
of the 1960s and thought it had inherited the mastery of the
universe simply by being young. And perhaps it had. So heady

were its early years that my generation has been wholly defined by them, till now; we were as sharply marked by rock and roll as our parents and grandparents were by the two world wars (the music let everyone be communicants, it let everyone join in and feel they were partaking together of this sacrament, which was youth). But as we come to the end of our time a different way of categorising us is beginning to manifest itself: we were the generation who, over the long course of our lives, saw the shadow fall across the face of the earth.

Let us set it out. Our world is under threat, as it has never been before, from a malady previous generations did not anticipate: the scale of the human enterprise. Down the centuries, in considering human affairs, our attention has been fixed on their direction, on the implausible, wondrous journey from the flint hand-axe to the moon, via literacy and medicine and the rule of law; gripped by the exhilarating course of the venture, we have not noticed its sheer dimensions creeping up on us. We have been the casual watchers of the waterlily pond, that celebrated pond where the lilies, barely noticeable at first, double in extent every day; they may take fifty days to cover half the surface, but we have not grasped the fact that to cover the remaining half, of course, then takes but a single day only.

This is the sudden headlong rush of exponential growth. It took us all by surprise. After the long unfolding of the human story, after all the millennia of history and of prehistory, it happened in a mere four decades, well within a single human lifetime, indeed within my own: between my teenage years and my middle years, between 1960 and 2000, the world's population doubled, from 3 to 6 billion. (Then it added another billion in the next decade, and will grow by a further 3 billion in the four decades to come.) And not only did the numbers mushroom, in the poorer countries especially; consumption exploded in the richer nations as they grew richer still and the baby boomers, the luckiest generation who ever lived, lapped it up;

and while population doubled, the world economy in the same period grew more than six times bigger. Looking back, this now seems much the most consequential historical event of the second half of the twentieth century, of more fundamental import even than the development and spread of nuclear weapons, or the retreat from empires, or the Arab–Israeli conflict, or the failure of the socialist project.

When did humans, creatures of the genus *Homo*, first begin to modify the world in a measurable way? Almost certainly when anatomically and behaviourally modern people, that is, members of the species *Homo sapiens*, emerged out of Africa some time perhaps around sixty thousand years ago, and began to spread eastwards across the world, to Asia, then down to Australasia, then back north-westwards into Europe, and finally over the Bering Strait land bridge from Siberia into the Americas. Formidably advanced through their possession of language, they – *we* – displaced and almost certainly annihilated the earlier species of humans which had spread out of Africa long before them, *Homo erectus* in Asia and the Neanderthals in Europe (who may not have possessed fully developed speech); and while they were at it, they visited a similar fate on the enormous animals which, over millions of years, had everywhere evolved as the top layer of the mammal and marsupial fauna which we still possess today. We do not accord much imagining to these vanished behemoths. We should. It was a massacre unparalleled. By the end of the Pleistocene, the long epoch of the ice ages, whole continental guilds of great beasts had been extirpated by humans, by the hunter-gatherers, such as the Australian megafauna with its two-tonne wombat, diprotodon, or the megafauna of South America with its colossal ground sloths whose fossils Darwin found, or the megafauna of Eurasia with its giant Irish elk whose ten-feet wide and ten-feet high antlers make you gasp in surprise when you encounter them in the atrium of the biological sciences department at the University of Durham.

No one really knows what happened, of course, and some palaeontologists believe changes in climate may have been responsible, but the most persuasive arguments strongly suggest that humans took them out; we did it. Twenty thousand, thirty, even forty thousand years ago, we were already transforming the world around us, we were destroying on a grand scale; and our populations were minuscule. What must be the effect, then, when not only has the technology for earth modification advanced, in our stirring journey upwards, from the hand-axe to the chainsaw, from the deer shoulder-blade to the bulldozer, from the fish-hook to the mile-long driftnet and from the throwing spear to the automatic rifle, but when we ourselves have undergone an upsurge in numbers which can only be described as gargantuan?

It is extraordinary: we are wrecking the earth, as burglars will sometimes wantonly wreck a house. It is a strange and terrible moment in history. We who ourselves depend upon it utterly are laying waste to the biosphere, the thin, planet-encircling envelope of life, rushing to degrade the atmosphere above and the ocean below and the soil at the centre and everything it supports; grabbing it, ripping it, scattering it, tearing at it, torching it, slashing at it, shitting on it. Already more than half the rainforests are gone, pesticide use has decimated wild flowers and the insect populations of farmland and rivers, the beds of the seas are deeply degraded and most of the fish stocks are at danger levels, the acidity of the ocean is steadily rising, coral reefs are under multiple assault, 40 billion tonnes of climate-changing carbon are loading the atmosphere every year and currently one-fifth, and rising, of all vertebrates – mammals, birds, fish, reptiles, and amphibians – are threatened with extinction. Many are on the brink, if not already gone. The Vietnamese rhinoceros was discovered in 1988, one of the thrilling secrets of the Indochinese jungles which war had for so long kept out of reach; it was extinct by 2010, slaughtered for its horn, believed in traditional Asian medicine, quite erroneously, to be a cancer cure. We knew the dodo for three times as long. The nightingale, the world's most versified bird, was revealed in 2010 to have declined in England by 90 per cent in forty years; that is, to have vanished from nine out of every ten sites where it sang as the Beatles were breaking up. The Mediterranean bluefin tuna, a fish glorious in form and function but unfortunately glorious too in taste, is starting to look doomed by the appetites of sushi eaters; all seven species of sea turtle are endangered, three of them critically; and amphibians are sliding in a bunch down the steep slope to oblivion, with the golden toad of the cloud forests of Costa Rica famous for its disappearance, while the golden frog of Panama may not be famous, but has disappeared just the same. Loss is everywhere, and the defining characteristic of the natural world in the twenty-first century is no longer beauty, nor riches, nor abundance, nor, if you like, life force, but has become vulnerability.

It cannot be stressed enough: these losses are not caused by natural events, such as tsunamis or volcanic eruptions. They are the work of people – of us – and as we continue to grow, and our needs continue to expand, so will the devastation. The proximate causes can be easily enumerated – we can see that they are habitat destruction, pollution, over-exploitation or over-hunting, the havoc caused by invasive species and, increasingly, a changing climate – but the ultimate cause of the great spreading ruination remains *Homo sapiens*: just one of the earth's great array of millions of radiated life forms, whose numbers, having exploded beyond the planet's ability to carry them, are now firmly on course to wreck it.

In a curious historical coincidence, at the very time when the explosion in numbers was beginning, a new vision of the earth it was so direly to affect was vouchsafed to us. We can put a precise date on it: Christmas Eve 1968. The person directly responsible was William Anders, an American astronaut, one of the crew of Apollo 8, the first manned spacecraft to leave the earth's orbit and circle the moon. When, on 24 December, he and fellow crewmen Frank Borman and Jim Lovell emerged in their craft from behind the moon's dark side, they saw in front of them an astounding sight: an exquisite blue sphere hanging in the blackness of space. The photograph Anders took of it is known as *Earthrise*, and its taking was without doubt one of the profoundest events in the history of human culture, for at this moment, for the first time, we saw ourselves from a distance, and the earth in its surrounding dark emptiness not only seemed impossibly beautiful but also impossibly fragile. Most of all, we could see clearly that it was finite. This does not appear to us on the earth's surface; the land or the sea stretches to the horizon, but there is always something beyond. However many horizons we cross, there's always another one waiting. Yet on glimpsing the planet from deep space, we saw not only the true wonder of its shimmering blue beauty, but also the true nature of its limits. Seen in the round, not really very big at all – the Apollo 8 astronauts could cover it with a thumbnail – and most assuredly, isolated. Only the one. Nowhere else for us to decamp to, in the never-ending blackness. Thanks to *Earthrise*, we now understand it in the intuitive way, in our souls: what we are wrecking is our home.

The idea that something might be done about this, that a way might be found to hold back the tide of human destruction across the globe, has been one of the great moral and intellectual challenges of the last quarter of a century, given that the pressures involved are intractable and that the problem itself is fully acknowledged by relatively few. They are usually classed as environmentalists or conservationists. They are in every country, and they are often loud, and sound influential, but they are small in number in global terms. Most ordinary individuals do not care, because the consequences are not yet visited upon them (although they will be), and also because people are quite naturally focused on their own concerns, which often seem harmless enough, and do not grasp that the essence of the trouble to come is their own individual choices, multiplied seven billion times.

Furthermore, the destruction of humanity's home by humanity's own actions is not something that can be coped with adequately – and that means, confronted – by our current belief system, which we might term liberal secular humanism. This creed, which has held sway since the Second World War, has a single, honourable aim: to advance human welfare. It wants people everywhere to be free from hunger and fear and disease, and in so far as is possible, to be happy and to live fulfilled lives. It is principled and upright. It is admirable. But there is a gap at its core: the failure to acknowledge that humans are not necessarily good. Still less does it admit that, more, there may be something intrinsically troubling about humans as a species: that *Homo sapiens* may be the earth's problem child.

Many, indeed, would be outraged by the suggestion, for poverty and hunger and disease are terrible enough without proposing that people as a whole are in some way flawed. Yet for the Greeks, the founders of our culture, this idea was central to their morality. There was a continual problem with man. Man was glorious, almost godlike, and continually striving upwards; yet only the gods were actually Up There, and if man tried to get too high, as he often did, the gods would destroy him. The gods represented man's limits. We think of Icarus, of course, but there are deeper lessons to be learned. The principal fault of Oedipus in Sophocles' *Oedipus Rex*, remember, was not that he murdered his father and married his mother; those were the incidentals of his fate. His real fault was that he thought he *knew* everything, he had answered the riddle of the Sphinx, he was beyond peradventure wise. The gods showed him he was not (and in the greatest of all tragic ironies, he puts out his eyes to punish himself for having been blind to his true situation, which now he can see).

In the modern consensus, in liberal secular humanism, this spiritual view of man as having limits, as not being able to do everything he chooses, and of potentially being a problem creature – for what else is a species which destroys its own home? – is missing entirely. There is no trace of it whatsoever. To suggest it, is absolute anathema. For with the dying of religion and the vanishing of spirituality we have become our own moral yardstick: at the heart of our notions of good and bad lies human suffering, and what we can do to avoid it. This is so deep-rooted in us now, so instinctive, that it has been internalised in the language: one of our most prized virtues is humanity, one of our deepest tributes to another person, that they are humane. He, or she, is a humane human. It's only one letter, one squiggle away from saying he, or she, is a human human. Our morality now is entirely anthropocentric: we automatically define objective good by what is best for ourselves. So where humanity's interests clash with other interests, the other are likely to get short shrift from us, even when they involve the proper functioning of the planet, which is the only place we have to live.

This has made the effective defence of the natural world very difficult in recent decades, especially in the face of development imperatives which may seem overwhelming. Environmentalists and conservationists, the people concerned with the fate of the earth's natural systems, have often been contemptuously dismissed by the development movement as middle-class birdwatchers, and it has long been hard to counter the assertive battle-cry of those turning rainforest, with its miraculous numbers of species, into nutrient-poor, soon to be exhausted farmland: *We Need It!*

What the defenders have tried to do, therefore, is construct a convincing response, to find an answer to the simplistic mantra of human necessity, which might bring to a halt the unthinking destruction of the natural world. There have been two serious attempts at this. The first has been the theory (or the project) of sustainable development. It has been a failure.

Mothered into the world by Gro Harlem Brundtland, some-time prime minister of Norway, via the 1987 United Nations' report on linking environmental and developmental concerns, 'Our Common Future', sustainable development seeks to let the mammoth human enterprise carry on growing, essentially to relieve poverty, without trashing its natural resource base. Sometimes referred to as 'green growth', it is officially 'development which meets the needs of the present, without compromising the ability of future generations to meet their own needs'. Again, this is admirable, and it can probably be done, as long as you think hard about it, and try; moreover, the theory accurately diagnoses both the problem and the potential solutions. The weakness is in the implementation. For sustainable development relies on the goodwill of people, and by extension, governments, to be put into practice; it relies on them changing their behaviour. It does not take into account that people are not necessarily good – and as such it was a perfect fit for liberal secular humanism, which does not take that into account either – and that people do not voluntarily change, if that means, stop acting out of self-interest. You might as well ask cats to stop chasing birds.

It would of course be unthinkably glib to dismiss out of hand the efforts of thousands of dedicated people, and the pursuit of sustainable development has made a real difference: above all, it has embedded, in governments and companies, the crucial idea, as a policy objective, that the environment must be taken into consideration, which was not there before. But what it has not done is alter fundamentally the general direction or the pace of the destruction of the natural world. It was thought that that might be possible when, at the Earth Summit in Rio de Janeiro in 1992, more than one hundred world leaders came together to endorse the theory and the gigantic work programme put together to implement it around the globe, Agenda 21; there was a moment of high hope and self-congratulation, as if drawing up the detailed solution to the problem were the same as carrying it out. I remember it vividly. I was there. But two decades later, in the follow-up conference, Rio+20, nothing was clearer than how complete, in terms of Saving The World, had been sustainable development's inability to deliver.

By 2012 little if anything had got better: with an additional 1.5 billion people added to the world, annual emissions of climate-changing carbon dioxide had increased by 36 per cent and were rushing upwards, another 600 million acres and more of primary forest had been chainsawed, pollution was soaring, especially in the developing world, and more species than ever were being threatened with extinction. Although there might have been successes at the margins, the main direction of destruction had not been diverted, and Rio+20, which convened in the Brazilian city once again and was the biggest meeting ever held by the UN – attended by 45,000 delegates, observers and journalists, including 130 heads of state and government – made a very weak, renewed commitment to sustainable development as a principle, and then was forgotten the instant it was over.

However, the second attempt at finding the answer is not yet a failure, and is currently sweeping the globe.

Sir Arthur Tansley is by no means a household name, certainly nothing like as familiar to us as his inventive contemporaries Ernest Rutherford, John Logie Baird and Alexander Fleming;

yet in the period of scientific ferment between the wars when all were active, Tansley, Professor of Botany at Oxford, conceived and popularised a concept which was to be just as influential as Rutherford's nuclear physics, Baird's television or Fleming's penicillin: it was the ecosystem.

It had taken natural scientists, obsessed with classifying things, a long time to realise that individual species of plants and animals do not exist in isolation, but in close communities formed with other living organisms, which all interact not only with each other but also with their surroundings; it was a perception not formalised, as the new science of ecology, until the beginning of the twentieth century. Tansley was one of the first prominent ecologists, and his introduction of the *ecosystem* term (in a 1935 paper devoted to an abstruse argument about ecological terminology) made graspable, even to non-specialists, the powerful idea of a living complex of animals and plants, working together with non-living parts of the environment such as the soil or the climate, as a functional unit.

Such units could be as large as a lake or as piddling as a puddle, they could be a forest or a single tree, but it was clear that they were real and they did indeed have functions, and in the sixties, seventies, and eighties, as they began to be studied intensively, biologists started to appreciate that they played major roles in modulating the way water, and nutrients, and sediments, and carbon all flowed through landscapes, from living things to the soil and the sea and the atmosphere and back again.

This understanding eventually crystallised into the even more pertinent perception that ecosystems and their associated wildlife did things for *us*, things which were vital: they provided life support services which we might always have taken for granted, but which we could not do without. Pollination of crops by bees and other insects is perhaps the most obvious example: without it, swathes of global agriculture would collapse. But by the 1990s scientists were starting to list more and more of these services: they included climate regulation, composition of the atmosphere, provision of fresh water, flood defence, control of erosion, maintenance of soil fertility, detoxification of pollutants, pest control, provision of fisheries, waste disposal, nutrient recycling, and more subtly, provision of a vast genetic library offering potentially life-saving new drugs and other products.

All that and more, we took from nature, without a thought. We had being doing so for aeons, because it was all free and so it was unnoticeable. The elucidation of the real role of ecosystem services, and even more, of our absolute dependence on them, has been one of the greatest breakthroughs in our understanding of the natural world, and what gave it peculiar force and relevance was that it came just as many of these services, for the first time in history, were under threat or actually being degraded.

Take the toppling rainforests. They could no longer be dismissed by their destroyers as mere pleasure gardens for bourgeois birdwatchers. Now we understood that they not only provided fuel and water and food, but also helped to regulate climate for us, and in a time when human carbon emissions were threatening to alter the atmosphere with disastrous consequences, they constituted a colossal carbon store which, many scientists and policymakers began to argue, it would be suicidal to sacrifice. (And their myriad plant species might very well hold an undiscovered substance which would save your child's life, like the rosy periwinkle from the forests of Madagascar, which gave us vincristine, a cure for childhood leukaemia.)

Our utter dependence on nature: here was nature's best possible defence, potentially far more effective than the hopeful pieties of sustainable development. The significance was seized on by conservationists, and the science of ecosystem services quickly grew into a discipline of its own: we might say it was formalised with the publication in 1997 of a compendium of essays entitled *Nature's Services: Societal Dependence on Natural Ecosystems,* edited by Gretchen Daily, a biologist at Stanford University in California. Since then it has exploded, being brought to popular attention globally by the United Nations with its Millennium Ecosystems Assessment, a vast survey published in 2005 which looked at twenty-four natural support systems for human life across the world, and asserted that at least fifteen of them were in serious decline. But our dependence upon them, vital though it may be, is not the only aspect of ecosystem services which has caught people's imagination. There is another perception abroad about nature which is exciting many: there's money in it.

Across the globe an extraordinary exercise is being carried out, one of the most remarkable society has ever undertaken: a great universal pricing. All over the planet, price tags are being affixed to grand chunks of nature, just as they are affixed to items on the shelves by a supermarket worker with a label gun, yet these are not the prices you might see on a can of beans or a packet of cornflakes, these are of a quite different order and say things like Pollination, 131 billion dollars, Coral Reefs, 375 billion dollars, Rainforests, 5 trillion dollars.

For the developing science of environmental economics has enabled us to accord ecosystem services value, real-world financial value, and this has woken up even more people than has the knowledge that we rely utterly on them. Take the example of mangroves, the salt-water woodlands found fringing many coastlines in the tropics. Imagine that the authorities in coastal zone X, with a rapidly expanding city behind it, decide to cut down its mangrove swamps because the shallow waters in which they are rooted provide an ideal site for shrimp farms, and if developed properly, those shrimp farms might produce, let us say for the sake of argument, 2 million dollars' worth of exports over five years.

But mangroves aren't just floppy trees with their feet in the water. They provide substantial natural protection against storms and tidal surges. Let us say that after the mangroves have gone, a tidal surge occurs, perhaps even a tsunami, which sweeps effortlessly over the shrimp farms and inundates the coastal region, and its city, to disastrous effect, and leaves the authorities of X with no alternative but to provide future protection by building a long sea wall. How much will the sea wall cost you? Say it's 200 million dollars, over five years.

The mangroves did it for nothing. So 200 million dollars is their replacement value.

And you got rid of 200 million dollars' worth of mangroves for 2 million dollars' worth of shrimp farms?

That sort of calculation has made people stop and think. It's the kind of arithmetic which can stay the chainsaws when *Please don't do it!* falls on resolutely deaf ears.

Environmental economists and many conservationists have been seized by it, and echoing in their minds is the biggest calculation of all, the sum worked out by a team led by Robert Costanza, then at the University of Maryland, which sought to put a financial value on all the principal natural systems of the planet which support human life. Published in the journal *Nature* on 15 May 1997, and attracting startled attention from all around the world, the paper by Costanza et al. estimated the central worth of seventeen of the earth's major ecosystem services at 33 trillion dollars annually – that's 33,000 billion, remember, 33 followed by twelve noughts – at a time when global GDP (all the goods and services produced by everybody in the world together) had an estimated annual value of 18 trillion dollars only. There it was. The value of nature to human society. Worth more than everything else put together. Nearly twice as much, in fact.

You can understand why many of those seeking to defend the natural world from destruction in the century to come, now see the economics as the answer; not least, it is far more aligned, than is sustainable development, with the hard-faced reality of the human condition as Adam Smith unforgettably expressed it: 'It is not from the benevolence of the butcher, the brewer or the baker that we expect our dinner, but from their regard to their own interest.' While sustainable development, alas, principally appeals to people's better natures, the concept of ecosystem services appeals to their self-interest directly. Follow the money, as Deep Throat told Bob Woodward.

Governments in the rich world have proceeded to do just this. As a sequel to the Millennium Ecosystems Assessment, they have been enthusiastic in setting up the TEEB project, a major global study of the economics of ecosystems and biodiversity, which reported convincingly in 2010 that saving the earth's wildlife from the crisis engulfing it would cost far less than letting it disappear (because replacing the services it provides would be unthinkably expensive) – the sort of statement which makes finance ministers, the people holding the real levers of power in many governments, everywhere sit up.

Yet just as I cannot be the only person who thought that something was lost to us when Neil Armstrong plonked his great fat boot down upon the moon, despite being in awe of the daring and the technological triumph (why? Because the mystery was no more), so I cannot be the only one who views these developments, powerful aids to saving the planet though they may be, with deep unease.

It is partly that the commodification of nature may strike many people as intensely unpleasant, not to say sinister: putting cash prices on rivers and mountains and forests is not a noble undertaking. To treat the elements of the natural world as commodities paves the way for them to be traded, speculated on, and ultimately owned and controlled by multinational corporations. The jargon of the financial world has begun to attach itself, and they cease to be places to delight in and become instead Natural Capital and Green Infrastructure.

But it isn't just that. It's even more, that the value which is accorded by the commodification of nature is highly selective. Worth is attributed only to services whose usefulness to us can be directly measured. For example, a recent innovative study suggested annual values for four ecosystem services in the USA: dung removal ($380m annually), crop pollination ($3.07bn), pest control ($4.49bn), and wildlife nutrition ($49.96bn). These values are based on human society trying to supply artificial replacements. But if it hasn't got a measurable, utilitarian value to us, it's nowhere, and by implication, not worth protecting.

For what value, in all this exciting new endeavour, do we give to butterflies, the creatures which, when I was seven, captured my soul? What value do we give, for that matter, to birdsong, which has captured countless spirits more? Are they just to be written off, as the great ruination of nature gathers pace? And the appearance of spring flowers or autumn mushrooms, and the unfolding of ferns, and the rising of trout, they have no value either do they, and is there now to be only one worth for wildlife, the one recognised by accountants?

Here we are at a peculiar moment in history, when the natural world is mortally threatened as never before, and those who love it are crying out for a defence. Yet while a new defence is being offered – one which is far more realistic and hard-headed than previous defences, one which must stand a better chance of succeeding – as we examine it, we realise that it too is deeply, crucially, fatally flawed.

What are we to do?

In a famous preface to one of his short novels, Joseph Conrad pointed out that the enterprise of the scientist or the intellectual may have more immediate impact, but that of the artist is more enduring because it goes far deeper; the statement of fact, however powerful, does not take hold like the image does. I believe that in defending the natural world, the time has come to offer up the images.

What I mean is, it is time for a different, formal defence of nature. We should offer up not just the notion of being sensible and responsible about it, which is sustainable development, nor the notion of its mammoth utilitarian and financial value, which is ecosystem services, but a third way, something different entirely: we should offer up what it means to our spirits; the love of it. We should offer up its joy.

This has been celebrated, of course, for centuries. But it has never been put forward as a formalised defence of the natural world, for two reasons. Firstly, because the mortal threat itself is not centuries old, but has arisen merely in the space of my own lifetime; and secondly, because the joy nature gives us cannot be quantified in a generalised way. We can generalise or, indeed, monetise the value of nature's services in satisfying our corporeal needs, since we all have broadly the same continuous requirement for food and shelter; but we have infinitely different longings for solace and understanding and delight. Their value is modulated, not through economic assessment, but through the personal experiences of individuals. So we cannot say – alas that we cannot – that birdsong, like coral reefs, is worth 375 billion dollars a year in economic terms, but we can say, each of us, that at this moment and at this place it was worth everything to me. Shelley did so with his skylark, and Keats with his nightingale, and Thomas Hardy with the skylark of Shelley, and Edward Thomas with his unknown bird, and Philip Larkin with his song thrush in a chilly spring garden, but we need to remake, remake, remake, not just rely on the poems of the past, we need to do it ourselves – proclaim these worths through our own experiences in the coming century of destruction, and proclaim them loudly, as the reason why nature must not go down.

It is only through specific personal experience that the case can be made, which is why I will offer mine. I will explore why, remarkably, we as humans may love the natural world from which we have emerged, when the otter does not love its river, as far as we know, and I will explore how it can offer us joy, through my own encounters with it over many years, touching on the ways it has touched me, just as it may have touched you; and I will do so, not just as a celebration of it, but as a conscious, engaged, act of defence. Defence through joy, if you like. For nature, as human society takes its wrecking ball to the planet, has never needed more defending.

2

Stumbling Upon Wilderness

The range of emotions which the natural world can excite in us is extensive, and we should not forget that it includes fears and even hatreds. Nature is not always benign. It can be dangerous. It can kill you. Some of the feelings it induces in us can be negative in the extreme (for example, wolves in the wild can arouse violent detestation in some – although reverence in others). But if we leave aside the blank-faced indifference which is also a major response to nature today, especially amongst young people seduced by screens and living electronic lives, then many if not most of the feelings it sparks in us are positive ones. Some we might characterise as satisfactions, such as the cherishing of familiar landscapes; others are sharper pleasures of novelty and beauty, such as encounters with rare and charismatic wildlife. A particularly powerful feeling is the sensation of wonder, which can whisper of immanence even to intensely practical personalities. But I am concerned with something which has preoccupied me more and more over the years, the most powerful feeling of all.

It is unusual. But it is not as uncommon as its exceptional character might suggest, and I have known not a few people who have encountered it. It is this: there can be occasions when we suddenly and involuntarily find ourselves loving the natural world with a startling intensity, in a burst of emotion which we may not fully understand, and the only word that seems to me to be appropriate for this feeling is *joy*, and when I talk of the joy we can find in nature, this is what I mean.

It seems to be an abrupt apprehension that there is something exceptional and extraordinary, beyond our everyday experience, about nature as a whole, something much more than the sum of its parts, glorious even though some of those parts may be, from birds of paradise to coral reefs, from Siberian tigers to bluebell woods. It is a sentiment that might be described as having a spiritual aspect, although it seems able to penetrate the most otherwise-secularised of minds. It can come to us anywhere, in the presence of a whole landscape or of a single organism; it can be met with in a number of nature's diverse aspects, such as its abundance or the peace it can bring; it can especially be encountered in the shifting calendar, in the discerning of a great change coming in the rhythms of the life of the earth, above all in the sensing of rebirth, in the springtime. A particular point: the wilder the part of the natural world you are in contact with, the more likely you are to experience it. I do not think it can be met with at second hand, through television wildlife documentaries, for example, however inspiring some of these may be.

It seems to be rarely talked about, perhaps because it is dimly perceived for what it is, perhaps because it is hardly ever articulated; and let us accept also that referring to it as joy may not facilitate its immediate comprehension either, not least because joy is not a concept, nor indeed a word, that we are entirely comfortable with, in the present age. The idea seems out of step with a time whose characteristic notes are mordant and mocking, and whose preferred emotion is irony. Joy hints at an unrestrained enthusiasm which may be thought uncool; to not a few in my country, it may seem merely old-fashioned and a bit lame, like patriotism. It reeks of the Romantic movement. Yet it is there. Being unfashionable has no effect on its existence.

If we examine what we traditionally mean by it, we realise that joy refers to an intense happiness, but one that is somehow set apart. It is in no wise the same as *fun*, or even *delight*, or, for that matter, terms describing extremes of gratification such as *bliss* or *rapture*, which in our sardonic age can no longer really be employed at all unless archly, in cookery writing, say (*new potatoes crushed with first cold pressing extra virgin olive oil – bliss!*). Joy, however, even if slightly uncomfortably, still lives in our lexicon with its original meaning, and what it denotes is a happiness with an overtone of something more, which we might term an elevated or, indeed, a spiritual quality.

We do not commonly employ it to define our pleasure, even if extreme, in eating a particularly well-made pork pie; but we might well think it appropriate to describe the feeling of a parent finding safe and well a missing child, or the feeling of a lover whose love for another person has long been unrequited, but who finds it at last being returned. It is not a word, I would say, that we use to describe self-centred gratification (and so we would tend not to use it for euphoria, even of the most potent kind, induced by drugs); it looks outwards, to another person, another purpose, another power. Joy has a component, if not of morality, then at least of seriousness. It signifies a happiness which is a serious business. And it seems to me the wholly appropriate name for the sudden passionate happiness which the natural world can occasionally trigger in us, which may well be the most serious business of all.

It is not exclusive, this emotion; it is not the property of the illuminati, of an enlightened or privileged few. It is open to every one of us, and I encountered it, purely by accident, when I was fifteen.

The memory is resonant still. The accident was one of geography, of having an unfamiliar and very different environment close enough to my home to be discovered, and what led me there was the pursuit of birds; for though it was indeed butterflies which first held sway over me – and they have continued to do so in a particular way, throughout my life – it was birds, once my enthusiasm had widened to all of nature, which led me out of childhood and into adolescence. In that, I was like a million other small boys in England in the 1950s (and judging by the membership figures of the RSPB, the Royal Society for the Protection of Birds, in subsequent decades, a million is probably not an exaggeration).

In fact, I think that in many ways I had a very typical fifties childhood, although this was strictly by dint of turning my back on the abnormality all around me, for after her breakdown of 1954 and her shaky and troubled return, Norah my mother, who clearly was only partially recovered, suffered two more collapses of her psyche, in 1956 and 1958. Both times were preceded by great distress within the family – I can remember cups of hot tea being thrown at the wall – and both required periods of recuperation in hospital for her lasting weeks, during which, with my father as ever away at sea, Mary and Gordon again looked after John and me, temporarily moving in to the new house my parents had bought, which was in another Bebington cul-de-sac called Norbury Close, about a mile from Sunny Bank.

For holding us together in the midst of all this, I realise now, Mary and Gordon were beyond praise. But even more than the first time, John could not bear it. He could not bear Norah's disappearance. He was driven hysterical with upset, quite literally: he began to scream in the street. He screamed at the prospect of her going away again, screamed after she had gone, and screamed that Jesus didn't love him, and the local children began to call him 'the mad boy' – I once actually heard two girls on rollerskates rattle past our front gate, one saying to the other, 'that's the house where the mad boy lives'.

Once again, I was indifferent. But the indifference was perhaps not quite so innocent now. For although there were clearly powerful psychological mechanisms holding sway over my feelings, something unsavoury had begun to creep in: a sense of embarrassment. As I got older I grew ashamed of John, and far from trying to assuage his distress, or even be his occasional comforter, I regarded him as an embarrassing encumbrance (it was the very opposite of 'He ain't heavy, he's my brother'; as far as I was concerned, he weighed a ton). I grew ashamed of my mother, too, for aspects of her behaviour which were bizarre and disturbing; and to this day I am torn between remorse at the way I deliberately turned away and did not support them, and anger that I should feel any such guilt, and talking about it in the formal setting, in the quiet room, trying to understand it all, long after both were dead, I suddenly burst out that I thought that they formed the Mad Club, which I was excluded from, and they excluded me because I wanted to be normal and *why was that wrong?*

So I turned my back on the acute distress of my mother and my brother; I resolutely looked away, pretended everything was fine, seized hold of normality and got on with life, and life became very much about nature, and birds in particular. Sixty years ago many young schoolboys in Britain, following a tradition no doubt centuries old, found birds a ready fascination, not least for their eggs, which were then being avidly collected (although the practice had just been formally made illegal, in 1954). I collected them myself, beshrew my soul, and I remember clearly a conversation between a group of nine-year-olds about the appearance of a wood-pigeon's egg, something quite unimaginable today. Lots of boys collected butterflies too, and even moths – scrawny little pipsqueaks were familiar with yellow underwings and cinnabars and five-spot burnets (although wild flowers were for girls) – but it was birds that were the bigger part of what we now call youth culture, perhaps as much as computer games are today. With natural abundance still flourishing, birds were all around us, even in the suburbs – every day the lawn and the rear hawthorn hedge at Norbury Close were swarming with house sparrows and tree sparrows, hedge sparrows and starlings, blue tits and great tits, blackbirds and song thrushes, robins and wrens – and so an interest in them was natural; but I imagine in most cases a real enthusiasm needed something specific to spark it, and in my own case it was tea cards.

Tea cards were the junior relatives of cigarette cards, those small portraits, usually at first of sporting celebrities such as cricketers in Britain and baseball players in America, which from the late nineteenth century onwards began to be given away in cigarette packets by tobacco companies as a marketing attraction. Forming collectable sets of twenty-five or fifty, they proved enormously popular, and eventually thousands of different sets were issued and the subject matter expanded far beyond sportsmen to everything from cars to cathedrals, from fishes to flags of the world. Gradually the marketing wheeze spread from cigarettes to other popular purchases such as, in America, chewing gum, and in Britain, tea. In the post-war years all Britain's major tea companies were giving cards away in their packets, the most energetic being Brooke Bond with its PG Tips (which eventually became the leading tea brand in the country thanks to its popular advertising campaign featuring chimpanzees dressed in human clothes and having a tea party, which half a century ago was thought of as a real rib-tickler).

Brooke Bond's cards were very popular, because the company not only produced numerous different sets of them but also went to the trouble of providing simple but appealing and extensively captioned albums to hold each set, attractively priced at 6d, or as we used to say in those pre-decimal currency days, sixpence (or, as we also used to say, a tanner). We took PG Tips in our house, and the cards in the set which caught my eye began to tumble out of the packets from 1958, when I was eleven. It was called Bird Portraits, it was a series of bird paintings by somebody called C. F. Tunnicliffe, and my interest was sparked immediately by the first one I laid eyes on: the stonechat. It was dazzling. The brightness, the colours, the *life* of the image, were magnificent. Here was a bird, perched atop a sprig of yellow flowering gorse, which was the very acme of alertness, displaying a black head and contrasting white collar with a breast of a wonderful fiery orange. It was a striking contrast not only with the birds on the lawn at Norbury Close, which, pleasing though they were, tended to be a tad on the toneless side, but also with the species I had been gazing at in *The Observer's Book of Birds*.

It confused and disappointed me, this book, especially after the youthful reverence I had accorded *The Observer's Book of Butterflies*. Both were early volumes in the long series of *Observer's* books produced by Frederick Warne & Co., the publishing house which had given Beatrix Potter and Peter Rabbit to the world. (*Birds* was the first, in 1937, and *Butterflies* the third, in 1938; *Wild Flowers* came in between.) They were pleasing pocket-sized handbooks on wildlife and hobbies and sports, on ferns and fungi, on coins and cricket, compact and inexpensive at five shillings each (or five bob, as we further used to say, in pre-decimal days); they were to delight three generations of British children, and adults as well, with the ninety-eighth and final title, *The Observer's Book of Opera*, appearing in 1982.

With a text written by Miss S. Vere Benson, a doughty campaigner who had founded the Bird Lover's League in 1923, *The Observer's Book of Birds* was fine as a mini-handbook, as a proto-field-guide, as far as factual information went; the problem was its illustrations, which were paintings. Some of them were excellent, even memorable, but some were rather weird, not seeming quite realistic (and there was no artist information). I did not understand why until years later when, in common with other nature-loving bibliophiles, I began to collect and research the series, many of which could still be picked up cheaply in junk shops, and I learned that Frederick Warne & Co. had owned the rights to numerous sets of wildlife illustrations from Victorian times onwards, and had simply reused them for the *Observer's* books.

The ones for the birds volume were taken from a celebrated and sumptuous collection of more than four hundred bird paintings, published from 1885 to 1897 in seven volumes and still sought after today, entitled *Coloured Figures of the Birds of the British Islands*, which was put together by the 4th Baron Lilford, a birding nobleman who was the long-time president of the British Ornithological Union (and the man who, on his estate in Northamptonshire, introduced into Britain the bird of the goddess Athena, the little owl). Lord Lilford brought three artists together for *Coloured Figures*: a Scot, Archibald Thorburn, and an Englishman, George Edward Lodge, both only twenty-six (the work was to make their reputations); and a Dutchman, John Gerrard Keulemans, who was in his forties. Thorburn went on to become one of the greatest of all bird artists, now residing without doubt on the upper slopes of that mountain on whose summit we place John James Audubon; Lodge is also greatly esteemed, if not to the level of his Scottish contemporary. The case of Keulemans is more problematic. An authoritative modern history of wildlife art refers to his output as 'workmanlike', and I eventually discovered that the paintings I found unconvincing in *The Observer's Book of Birds*, at the age of nine or ten, came from his brush. They were just not true to life. It troubled me. Take the yellow wagtail. In the flesh it is the epitome of the dainty. In Keulemans' image, it was a great fat lump. The pied wagtail was no better. But the one which really confused me was the willow warbler, which is the size of a man's thumb or smaller, but which, thanks to John Gerrard Keulemans' interpretation, I was unshakeably convinced, for an inordinate amount of my young childhood, was two and a half feet long.

There was nothing confusing about Tunnicliffe's willow warber. It fell out of the tea packet neat and tight and slick, like the real bird; he had caught it, on a branch of willow catkins, in mid movement, in the jerky incessant gleaning of tiny insects off twigs and leaves. (And seen *from above*, which I now think is just so clever and original.) As the McCarthy family, like the rest of Britain, lapped up its PG Tips, image after image came out of the packets to surprise and delight (and of course, to be collected). They were instantly striking, not only for the vibrancy of their colour, which was often exquisite, but also for the boldness of their presentation: since the small size of the cards meant the space available was very limited, Tunnicliffe made the image fill the frame and often it did so dramatically, the gannet plunging, the teal springing, the barn owl pouncing, the sedge warbler singing, the pied flycatcher spreading its wings in a sort of flycatching ecstasy. Some images, like the bullfinch surrounded by apple blossom, the goldfinch amidst thistledown, or the water rail emerging cautiously from the vegetation next to a clump of marsh marigolds, were calmer, but breathtaking in their beauty. They were charming miniatures, every one.

Today, of course, I am fully aware of Charles Tunnicliffe and his achievement. I share the opinion of many that he is the pre-eminent British bird artist of the mid twentieth century, and certainly, his *Shorelands Summer Diary*, filled with sketches as well as paintings of the wildlife around his home in Anglesey, is one of the loveliest books on the natural world ever produced. But back then, at the age of eleven, I had no notion of artistic achievement. I neither knew nor cared who C. F. Tunnicliffe was. All I cared about was this pageant of marvellous birds that he was producing and I was collecting: I wanted to see them properly, to set eyes on them for myself in real life, and I began to search for them, on foot and on the bike I had been given for my eleventh birthday, in the lanes and fields and woods of the Wirral.

Birkenhead, the town where I was born, and Bebington, its suburb where I grew up, not only sit across the River Mersey from Liverpool; they sit on a peninsula. Fifteen miles long by seven wide, the Wirral has a river on each side and the sea at its end: the Mersey to its east, the River Dee to the west, and the Irish Sea to the north. When I was a boy, it had since time immemorial been part of Cheshire, and at its base was Chester, Cheshire's county town, where nearly two millennia ago, I was proud to learn, the Twentieth Legion, *Valeria Victrix*, the Conquering Valerian, with its badge of a wild boar, built their camp on the Dee, on the right bank of the river, which they called Deva. (They were there for more than three hundred and fifty years until the day came, probably in AD 410 with Rome collapsing, that they just upped and left, and the American poet Stephen Vincent Benét, Hemingway's contemporary – he of the wonderful 'I have fallen in love with American names' – wrote a chilling short story imagining the legion's final departure from Chester and its march south, as its world began to disintegrate.)

History touched the Wirral several times after the Romans decamped, most notably in AD 937 when Æthelstan, the first king of England, defeated the combined army of the Vikings and the Scots in the greatest clash of arms the Anglo-Saxon world ever saw (before the fateful encounter at Hastings): the Battle of Brunanburh, which was almost certainly fought at Bromborough, a part of Bebington. In the fourteenth century the poet of *Sir Gawain and the Green Knight* had Gawain ride through 'the wilderness of Wirral'; in the eighteenth century Emma Hamilton, known to history as Nelson's squeeze, grew up in the Wirral village of Ness. But it was the nineteenth century which made the greatest mark on the peninsula, as its eastern, Mersey edge became heavily industrialised and Birkenhead, in 1800 just a small village with a ruined priory, swelled to become a major shipbuilding town, with by 1900 a population of 110,000 (and more than 140,000 when I was born, half a century on).

This development gave my Wirral its character. It made it somewhere, like the head of Kipling's Kim, with two separate sides; it intensified the difference natural geography already provided in the two rivers, which are quite opposite in temper. The Mersey on the east narrows to a bottleneck between Liverpool and Birkenhead, so it is full of deep water, the reason for Liverpool being one of the world's premier ports; the Dee, however, on the west, opens out into a giant funnel-shaped estuary of salt-marshes and mudflats. Nineteenth-century industrialisation merely exacerbated this contrast. The eastern side of the Wirral, my side, became urban, scruffy, cramped and impoverished, looking naturally across the Mersey's ship-jostling waters to Liverpool's famous cityscape, to its docks and factories and terraced streets; but the western side, looking out across the Dee and its estuary to the mountains of Wales, remained rural, unspoiled, desirable and affluent, studded with pretty sandstone villages like Burton and Parkgate and Caldy. The distinction persists, as a striking piece of symbolic social geography, and I am perennially surprised that no one seems yet to have written the defining Wirral novel of Terry, shall we say, the working-class lad from the Mersey side of the peninsula, falling in love with Tamsin, the upper-class girl from the Dee side; perhaps I have missed it. But in my own life, I was indeed drawn as a teenager from the one side to the other, although not for social or romantic reasons.

Bird fervour led me there. The estuary of the Dee is Elysium for the birdwatcher, a sprawling watery plain ten miles long and six miles wide at its mouth, holding, in winter especially, enor-mous numbers of waterbirds, both waders and wildfowl. As I walked and cycled around the Wirral, spurred on by C. F. Tunnicliffe and his tea cards, and my knowledge grew, and I began to spot the tree-creepers and siskins of Storeton Woods, and the yellowhammers and linnets of the fields beyond them, and the pheasants and partridges that ran in the fields, and the spotted flycatchers of various Wirral village gardens, I began to realise, talking to similar-minded boys, that the Dee had birds which were bigger and wilder and more exciting still, and that it was there that I needed to be. This was ultimately dependent, however, on a rite of passage common to all the birdwatching boys of my generation: the purchase of a pair of binoculars. They seemed to be very dear in those days, did bins. At Christmas 1960, therefore, when I was thirteen and a half, I asked for money for a binoculars fund instead of traditional presents: £2 was forthcoming, and duly put aside. It was complemented by another £2 for my fourteenth birthday in June 1961; a third such sum the Christmas afterwards; and finally, £3 for the birthday which followed, when I was able to buy, for eight pounds ten shillings, a pair of 8 × 32 field glasses of no famous make, but which at least were serviceable. And so, with them slung around my neck, in the summer of 1962, when I was fifteen, I walked out into the Dee estuary looking for birds, and what I stumbled upon was wilderness.

The idea that wild places, those which remain wholly untouched by people, might be of value to us and even cherished and protected, rather than just being thought of as waste land or worse, is relatively recent, in historical terms. Once, of course, there were no wild places as such; during the fifty thousand generations and more we spent evolving through the Pleistocene, the epoch of the ice ages, as hunter-gatherers, when we ourselves were an organic part of the natural world, all places were by definition wild; but then, in the greatest of all human revolutions, as the last of the ice disappeared about twelve thousand years ago, farming was invented. With the cultivation of crops and the domestication of animals, agriculture for the first time permitted stable settlements, it allowed for villages to be created, then towns, then cities and the rise of everything we call civilisation; but even more than that, it fundamentally altered our relationship with nature, from one of partnership, more or less – for even as hunter-gatherers we could be demanding partners – to one of formalised mastery and domination. We have spent most of the five hundred generations since, the Holocene epoch, breaking the sod and hacking the forest down and proclaiming our God-given right to do so, God-given quite literally – the Old Testament spelt out bluntly the farmers' ascendancy over nature, and their entitlement to do whatever they damn well liked with it, in the famous lines of Genesis, 1:28: 'and God said unto them, be fruitful, and multiply, and replenish the earth, and subdue it: and have dominion over the fish of the sea, and over the fowl of the air, and over every living thing that moveth upon the earth'. Thus we long regarded wild places and wilderness, the bits we hadn't managed to subdue or have dominion over, with near universal disapproval, indeed with a revulsion sometimes verging on horror. It was against wild places, after all, that the civilising struggle was being waged, to clear the forest and grow corn in its stead; the forest was the enemy, it held deadly wild beasts, and sometimes deadly wild men, as deserts did, or mountains. The civilised looked to the cities. What was there in wilderness other than the absence of everything that made life worth living? For aeons it was hated and feared and despised.

The shift in opinion that started to change this attitude, in the early 1700s, was fairly shallowly based: it was aesthetic. But it was effective, nonetheless. It began when English gentlemen started taking the Grand Tour of Europe, in the course of which they survived the vertiginous crossing of the Alps, and enjoyed having been terrified. So arose the influential concept of the Sublime, the appreciation of the awe-inspiring side to nature, something which was not quite the same as beauty but which prompted admiration just as powerfully. It became an influential literary and artistic fashion, and in the second half of the eighteenth century it was joined by another, slightly tamer vogue for viewing the natural world, wild places and all, in a positive artistic light, the concept of the Picturesque. Their combined influence meant that by the 1780s, especially as turnpike roads were built and public transport improved, Britain's once-despised wild scenery, on the River Wye in Wales and in the Lake District in England in particular, was attracting an increasing number of tourists, while in continental Europe, Jean-Jacques Rousseau had sung the praises of the Alps and insisted on the innate goodness of the natural world and of man himself; and all of these streams of thought fed into the swelling river that was Romanticism, until as the nineteenth century began, William Wordsworth could proclaim himself, from first to last, a follower of nature, and he himself was followed by many more.

So nature finally found its champions; but they were not specifically champions of wilderness, of what we might call wholly untouched, *unhumanised* land. Much of Wordsworth's Lake District, mountainous and awe-inspiring though it might be, was a farmed landscape, one way or another; it had people in it. It had Michael. It had Lucy. It could not really be called a wilderness. The champions of wilderness proper began to emerge fifty years later, in America.

It was only natural. In the United States, in this new-found world, the extent of completely untouched land was colossal, especially in the centre and west of the continent, which had never been settled. Wilderness was virtually the country's defining landscape. Yet despite its magnitude, it began to come under mortal threat as the nineteenth century progressed and the young country stormed headlong into the swiftest and most extensive mastering of nature the world had ever seen, breaking the sod and hacking the forest down on a continent-wide scale in just a few decades, as part of the westward expansion of the Frontier, an enterprise regarded by Americans themselves as so heroic that it came to symbolise for them their national character, with its virtues of individualism, self-reliance and independence. Year after year, the pioneers pushed further westwards and built their log cabins; the untouched prairies were ploughed; the ancient trees were toppled in their thousands; the indigenous inhabitants, the Native Americans, were evicted from their ancestral lands; and cattle replaced the buffalo herds, and the bear and the lynx and the wolf, all to the approbation of the citizenry as a whole.

And yet . . . even as this was all going on, doubts about the wisdom of so forcibly taming, often in effect destroying, the extraordinary wild landscapes which were being discovered in the west, many of them more freshly magnificent than anything in Europe, were growing in the minds of young America's own nature writers, led by Ralph Waldo Emerson and Henry David Thoreau, both of whom, as Transcendentalists, saw the unspoiled natural world as a way to spiritual truth. Thoreau went further and − perhaps the first person to do so − specifically championed the concept of *wildness/wilderness*. Although best-known for *Walden*, an account of two years living in a cabin in the woods, his forceful views on wildness are set out in *Walking*, a lecture given several times and published after his death in 1862, which contains the famous line 'In Wildness is the preservation of the World.' Thoreau saw man as being a part of nature, and he saw wild places not only as essential to human well-being, but also as a source of primitive strength; it was 'not a meaningless fable', he said, that Romulus and Remus, the founders of Rome, had been suckled by a she-wolf.

His support for wilderness was soon echoed by one of nineteenth-century America's most noteworthy public men, George Perkins Marsh, who is barely known in Britain, an omission that needs to be rectified. A lawyer, politician, diplomat, and outstanding linguist, successively US envoy to the Ottoman Empire and to Italy (where he died), a polymath and universal man who was almost the Victorian-era equivalent of Thomas Jefferson, Marsh was also an ecologist *avant la lettre*; and in 1864 he produced a book which is the first-ever summary of the *ecological* consequences of doing what Genesis urged us to do, and of dominating and subduing the earth. American critics not infrequently link it with Darwin's *The Origin of Species*, published just five years earlier, and while Marsh does not, like Darwin, overturn all previous conceptions of what humanity is and cannot be made the Englishman's intellectual twin, there is no doubt that in challenging another enormous assumption which people had always made about the world, his originality is of a comparable order.

The assumption was that doing things to the earth had no cost. It followed on inevitably from the Bible's declaration that the planet's resources were put there by God for our use, and thus by implication were boundless. That was woven inextricably into the Christian mindset. It is hard to overstate how fundamental, in a still Christian world, was Marsh's contesting of it. But in *Man and Nature (Or, Physical Geography as Modified by Human Action)* he did that at length, highlighting the early Mediterranean societies, which, he said, had collapsed because the deforestation in which they had engaged had destroyed their water supplies. He bolstered his case with his own vast travel experience and mammoth erudition – *Man and Nature* is a dense read – and moved inexorably to his point, which was that in its headlong conquest of the Frontier, America was in grave danger of repeating the mistakes that these earlier societies had made, and ruining itself.

His was the first voice to enunciate these insights, now commonplace amongst us. And he went further. Such was the sweep of his learning and the depth of his vision that he felt able to generalise about the baleful influence of the human species on nature as a whole, as he perceived it, and he did so in words as darkly memorable about us as Adam Smith's hard-headed remark as to why the butcher, the brewer and the baker provide us with our dinner. 'Man is everywhere a disturbing agent,' Marsh wrote. 'Wherever he plants his foot, the harmonies of nature are turned to discords.'

There, was the true value of wilderness, of unhumanised land: it was where the harmonies of nature, the balance and beauty of the natural world, remained. This was a far profounder assessment of its worth than the fact that it could give a gentleman a fright, and it became the intellectual underpinning of the devotion to wilderness which began to gain an increasing foothold in American thinking about the natural world. But it was not Marsh who orchestrated it. That was a torch taken up by John Muir, the Scots-born writer who emigrated to the United States aged eleven, in 1849. Muir spent his adolescence on his father's farm in the wilds of the Wisconsin frontier, and after an accident in which he nearly lost his sight, he realised that the wilds were where he wanted to spend his life. In 1868 he moved to California and discovered the mountains of the Sierra Nevada, a wilderness supreme, and for the next forty years and more he informed a growing audience of their transcendental qualities and why they mattered, in lyrical and sometimes quasi-mystical terms: undisturbed nature, he said, was 'a window opening into heaven, a mirror reflecting the Creator'.

By the end of the nineteenth century, then, the value of wilderness, something barely recognised in any other society, was in America formally and widely acknowledged, and the word itself, for long in use disparagingly – think of Jesus in the wilderness – was for the first time being used in a positive way. Thoreau, Marsh, and Muir had all seen something in wholly wild land which made the most powerful appeal to the human spirit, and their perception was increasingly shared. Muir became a national celebrity, not only for his writings but also for his wilderness activism, helping to bring about the creation of California's Yosemite National Park in 1890 and becoming the founding president of the United States' first major conservation body, the Sierra Club. By the time of his death, in 1914, the love of wilderness was becoming ineradicably established in the American mind, and as the new century went on it only grew, supported by thinkers such as the lyrical forester-philosopher Aldo Leopold, who called for a new 'land ethic' of ecological responsibility. It reached its climax in 1964 when President Lyndon Johnson signed the Wilderness Act, a piece of legislation establishing a National Wilderness Preservation System for America, a gigantic protection scheme for vast areas of untouched, unhumanised country, quite unlike anything else in the world.

But that was America. They could love their wilderness, because they had it. In Britain, although we cherish our countryside and its gentle beauty, and strive to protect it just as much, it is a landscape which has been farmed time out of mind; there is little that can justly be given the wilderness label, at least in its southern half, in the English lowlands. After leaving King Arthur's round table in his quest for the mysterious Green Knight, Sir Gawain might have ridden through 'the wilderness of Wirral' – *few thereabouts that either God or man with good heart loved* – yet that was written some six hundred years ago, and by the time I came along, Gawain's godless Arthurian wilderness was industrial town and suburb: it was Sunny Bank and Norbury Close. It was long tamed. On the Wirral's eastern side, anyway.

The Mersey side.

Where I grew up.

But the western side, the Dee side . . . well. Funny. It was not quite so clear-cut. I don't mean the gentle farmland with its oak-dotted hedges and red-brown sandstone walls, the pretty villages of Caldy and Parkgate and Burton, but the estuary . . . when you first see it, when you come round the corner on to the Parkgate promenade, say, and it's there smack bang in front of you, mile after mile of empty marshland stretching away uninterrupted to the Welsh mountains on its far edge and the sea at the far end . . . you are given pause. There's a definite feeling of immensity facing you, of nature untouched on the grand scale, which is hard to ignore.

Not that I registered it, when I walked out on to the Dee that summer with my new bins slung proudly around my neck and my spirit still in thrall to the scintillating images of Charles Tunnicliffe: I saw it simply as a bird area, an ornithological extension of the Wirral itself. The bottom, southern half of it was saltmarsh; the top, northern half, where the estuary met the sea, consisted of the intertidal zone, mudflats and sandbanks daily covered and uncovered by the tide. Initially I explored the saltmarsh edge, as that was nearer to where I lived, and I found lapwings and kestrels, skylarks and meadow pipits, herons and reed buntings; but I soon realised that there was even more happening at the estuary's mouth, at West Kirby and Hoylake, where flocks of wild duck, and especially waders such as ringed plovers, redshanks, oystercatchers, and curlews, fed and roosted on the bare mud and sand, but were pushed off by each incoming tide and so were active and very visible.

There were thousands of them: the Dee was overflowing with life. And the more I watched, the more I came to feel, as I still feel today, that the birds which live where the land meets the

49

sea are among the most alluring of all God's creatures. *Waders* is the English word, which describes their method of motion; Americans call them *shorebirds*, referring to their habitat. It is a useful term I will also sometimes use. Spindly-legged, nervy, refined, they epitomise elegance on the one hand, and on the other, wildness: they will not come to your garden, sit on your fence, hop on your lawn or sing for their supper; they remain in their own wild places, eternally untameable.

Yet at the heart of their existence, and of our feelings towards them, is a paradox. They are the gift to us of mud. Mud we find repellent, a substance a step away from shit; but the inter-tidal ooze at the edge of the sea is the richest in invertebrates of all habitats, able to hold in a single square metre thousands of tiny molluscs, crustaceans, marine snails, and marine worms, and waders are linked to it inextricably, having evolved to feed on it, indeed, to divide it all between themselves. The term in ecology is '*niche partitioning*': different shorebird species take different invertebrates from different places, and the main differ-entiation mechanism is bill length. Short-billed birds such as ringed plovers take organisms on the surface; medium-billed species such as redshanks start to probe into the mud for small gastropods; longer-billed oystercatchers probe deeper still, able to find cockles; and curlews with their decurved beaks, the longest of the lot, can find lugworms and ragworms at the bottom of their burrows. But all of them are united by a feat impossible for people: in moving over mud and slime and goo, they are never less than graceful.

They have something else about them to attract free spirits: they are world-wanderers. Many species in shorebird families such as sandpipers and plovers are highly migratory, journeying every spring to the High Arctic. From around the globe – not only from Europe, but also from Asia, Australia, and the Americas – they head for the far north, to the tundra at the top of the world, which in its brief but bountiful summer, with insect superabundance, extended daylight in which to feed, and relatively few predators, is a superlative place to fledge their chicks. They then return to spend the winter in mid-latitudes such as Britain, or push further south into the tropics, even penetrating deep into the southern hemisphere; and on the Dee, the end of the summer brought to the tidal flats, the mud and the sand, a great influx from the north. I encountered for the first time Arctic-breeding birds in their winter plumage such as sanderlings, grey plovers, greenshanks, turnstones, curlew sandpipers, dunlin, and above all the knot, the medium-sized sandpipers which formed immense flocks of tens of thousands of individuals, so colossal that when I first saw them in a shape-shifting dark murmuration, far in the distance, I thought I was looking at a billowing cloud of smoke, and wondered how big the fire must be.

But gradually I became aware of more than the birds. I started to become conscious of the place, of the estuary itself. You could not but be affected by it, if you spent time there. It was a realm apart. Like the waders themselves, it was wholly wild and untamed, even though it was a mere six miles from my home, in a suburb on the edge of a major industrial city. The sheer size of it was its most imposing aspect, especially if the sort of open spaces you were used to in your suburban existence were football fields or slightly larger municipal parks with bandstands and railings, litter bins and stern notices about dogs. This estuary too was a defined open space, but it was about 13,000 hectares in extent, or 35,000 acres, or 10,000 football pitches, and from one shore to the other, it was entirely devoid of human artefacts, being simply saltmarsh, sandbanks, and mudflats.

There was something more than its size, though, which added to the estuary's appeal to me. It sat on the shoulder of Wales. From the Cheshire side you looked across to Flintshire, which was in a different country, a nation with its own language and history and mountainous bearing (in stark contrast to the horizontal tranquillity of the Cheshire plain) – a country of otherness for which I had already conceived a deep attachment that has lasted all my life, and the fact that its slopes and summits were what you saw when you looked out over the Dee, for me, was spine-tingling: all the way down the estuary you could see the ramparts of the Flintshire hills, and behind them you could glimpse the tops of the Clwydians, the first mountain range; and if you went to the estuary's mouth at West Kirby and Hoylake, on some days you could catch sight of Snowdonia itself, you could see the Carneddau, Carnedd Llewellyn and Carnedd Dafydd, shadowy peaks in a dim and distant land.

I began to appreciate it all properly in September when I started exploring at the other end, at the head of the estuary, where it began, at a sandstone outcrop called Burton Point. Not far beyond it, at Shotton, was heavy industry, the giant steelworks of John Summers & Sons, but somehow this didn't detract from the landscape, and in fact, surrounding the steelworks were Shotton pools, a group of man-made lakes which formed a major birding site. I wrote to John Summers and they sent me a birdwatching pass giving me access to the pools, and to reach them I rode to Burton Point, hid my bike among the rocks, and walked the length of a mile-long embankment.

On one side of the embankment was an army rifle range; on the other, the estuary of the Dee. You were at its base, and you could turn outwards and view the whole of it, with Wales and its mountains on your left, the Wirral on your right, and the immensity of the estuary in between stretching to the level horizon, with its hint of infinity – that was the sea, more than ten miles away – and the great open skies. It was a very isolated and solitary spot (I never saw another soul there). I had gone looking for birds, and I had stumbled upon wilderness, as near as you will find it in the lowlands of England. I started to sense then the specialness of it all, it started to stir other parts of me; what brought it to a climax was music.

It was the music of the waders. I had come to know their

calls and come to love them. The commonest was the piping of the oystercatchers, most often a forceful *peep!*, which had an anxious air about it. I was also strongly drawn to the triple call of the greenshanks, *tew-tew-tew*, and even more to the two different sounds of the curlews, the sharp, carrying *cour-LEE* call, and then the strange melancholy bubbling song, which Dylan Thomas evokes in the Prologue to his 1952 *Collected Poems*, apostrophising 'the curlew herd':

> Ho, hullaballoing clan
> Agape, with woe
> In your beaks . . .

It's hard to remain unmoved when curlews are bubbling – it's a sound which àlters the landscape, especially in the spring – and I think the birds had an added mystique for me because when I was much younger, I had read and been captivated by Eleanor Farjeon's fairy tale *The Silver Curlew*, her reworking of the Rumpelstiltskin legend, set in Norfolk when Norfolk had a king, and I had ever after felt that curlews were creatures set apart. But it was another bird that moved me most.

It was a sandpiper, the redshank; and it had a call which, unusually, the bird book then used by every birdwatcher seemed to have set down accurately. I say unusually, because transcribing bird calls into human sounds is very much an inexact science, but *The Field Guide to the Birds of Britain and Europe*, by Roger Tory Peterson, Guy Mountfort, and P. A. D. Hollom, had with the redshank pretty much got it right.

'Usual call,' it said, 'a musical, down-slurred, *tleu-hu-hu.*'

I thought it was amusing to see it written down baldly like that, in consonants and vowels. *Tleu-hu-hu*: it could be a verb from an exotic language. But it did convey fairly closely the lilting, mournful sound the birds gave when they took flight, which could be borne far over the marshes on the wind, and which I found touched me more than anything else – finding that for sure one day which I know was in October, although I kept no specific record of the date.

October 1962 saw several great events full of contingent influence upon my life. There was the Cuba crisis, the cold war's most perilous nuclear stand-off, when I lay on the floor of the bathroom saying rosaries like other people were smoking cigarettes, and begging God to save us – no one not alive then can imagine the terror of that week – and the Second Vatican Council, opened in Rome by Pope John XXIII, Papa Giovanni, who began a rethinking of the severe creed in which I was being brought up, that eventually led me to rethink it myself – and the release of a first record, *Love Me Do*, by a local rock band from across the Mersey (although in those days we didn't call them rock bands, we called them beat groups) whose name was The Beatles, a record which by December had reached Number 17 in the national charts – something I remember chattering about excitedly at the school Christmas fair.

You might say it was the month that the sixties began, October 1962, when the great gates of change began to creak open. My own significant event from it holds no significance for anyone other than me, but it does still resonate with me strongly. I remember about the day itself, that I saw a goldeneye first. I had biked to Burton Point and started trudging down the embankment to Shotton pools and halfway along, I slipped down to the marsh itself so that I would not be silhouetted as I reached the pools and could come back up with stealth; and when I eventually did, and peered over the embankment top, there was a real prize: not fifty yards away on the water was the goldeneye, a splendid duck from Scandinavia which I had never seen before but recognised at once from the *Field Guide*.

I spent, I suppose, about an hour watching the pools and then headed back, and the weather was somewhat unusual for Britain: sunny, with a stiff breeze. The whole of the Dee was on my left hand, at peace in the golden light of October, and I began to hear faint sounds: redshank calls. *Tleu-hu-hu*. The birds were calling from somewhere invisible to me, out on the marshes, but their voices were being carried on the north-west wind which was blowing straight down the estuary's whole length towards me, and looking at it all, I stopped, sat down on the embankment and listened, and another call drifted to my ears, and it suddenly seemed to be pulling everything together, this ethereal mournful fluting, all the beauty of the untouched estuary and the great skies and the distant mountains, all its richness of life, and I realised for the first time where it was coming from: from the very heart of wildness.

Whatever it was that had captured the spirits of Thoreau and his successors, looking on the untouched landscapes of nineteenth-century America, in that moment on the Dee estuary captured mine. I saw a part of the earth in a way I had never seen it before. Or perhaps, I saw it with a different part of me.

Before, had you asked me about it, I would have said that the estuary was broad. It was long. It was flat. It was green. Or, sometimes, it was wet.

Now I would say something different: it was wonderful.

I loved it with as intense a love as I had ever experienced, and there, sitting on the embankment, in the sunshine and the wind, with the wild calls drifting to my ears, I looked on the natural world, and I felt joy.

3

The Bond and the Losses

So many powerful minds have addressed it, the unrelenting destruction of nature around the globe, so many experts have looked at the economics and the ecology and tried to reconcile them, so many thousands of detailed policies have been worked out and applied, so much intellectual effort and so much idealistic concern have been thrown at the problem, year after year after year, that the question presents itself at once: how on earth might it be the basis of a better defence, a better defence of the natural world, the fact that one autumn afternoon, more than half a century ago, a teenager sat looking down an estuary and suddenly felt happy?

We think of ourselves, especially since the decline of Christianity in the West, and its replacement by our current creed, liberal secular humanism, as rational beings entirely; we pride ourselves that, faced with a Problem, with a capital P, we may employ Reason, with a capital R, and naturally find a Solution, with a capital S. We believe that this will deliver, every time. Rationality is ingrained in a million mindsets. Yet the world does not always work like that (as those who lived through the two world wars, mired in chaos and evil, knew only too well). And there is another way of going about things, in dealing with the mortal threats that our planet now faces, which is to consider, not what we do, but who we are.

Most of us probably think we know. We do not give it a second thought. But in the last thirty years or so, a new understanding, by no means yet widespread or popularised, has begun to dawn of what it means to be human, based on a simple but monumental perception: the fifty thousand generations through which we evolved as hunter-gatherers are more important to our psychological make-up, even today, than the five hundred generations we have spent since agriculture began and with it, civilisation. We possess the culture of the farmers, the subduers of nature, and the citizens who came after with their settled lives and their writing and law and architecture and money, yes of course we do, but deep down, beneath culture in the realms of instinct, at the profoundest levels of our psyche − the new vision has it − we remain the children of the Pleistocene, the million years-plus of the great glaciations, when the natural world was not subdued and we lived as an integral part of it, in coming to be what we are. The legacy inside us has not been lost, and in many ways it is controlling.

The insight is from evolutionary biology, which in recent decades has moved on from exploring how, through Darwin's principle of natural selection, the peacock evolved its resplendent tail and the parrot its formidable beak, to looking at how in just the same way people evolved to be people; specifically, it is from the relatively new discipline of evolutionary psychology, which examines the ways in which the human mind adapted itself to the issues that Pleistocene hunter-gatherers faced in their daily lives, as over thousands of generations they gradually evolved inherent traits and instinctive reactions which remain with us still. The consequent account of what appears to be psychologically 'hard-wired' inside us, the list of putative human universals, is long and fascinating, from our fondness for sweet foods to our fear of snakes and spiders, from children's enjoyment of hiding to their predisposition to climb trees, from our ability to throw objects precisely at a target (which no other creature can do) to our pleasure in bodily adornment, from men's attraction to slim-waisted women (who appear to be not pregnant and thus available for mating) to women's attraction to high-status men (who can better defend them) – even to our preference for certain types of landscape.

And there, it gets more fascinating still. Surveys have demonstrated that, shown different landscape images, people overwhelmingly favour one form in particular, one of open grassland interspersed with trees and a view to the horizon, and if possible water, and animal and bird life; and it has been suggested that this closely resembles the tropical African savannas on which *Homo sapiens* evolved, before spreading out across the rest of the world. (The idea is known as the savanna hypothesis.) The reason that many thousands of years ago we might have developed attachments to certain landscape features so powerful that they became hard-wired in our genes and are with us today, is simple: it was necessary for survival. The hunter-gatherers of the Pleistocene were constantly on the move – their existence was memorably characterised by Gordon Orians, originator of the savanna hypothesis, as 'a camping trip that lasts a lifetime' – and choosing which new landscapes to enter and which to avoid must have been an absolutely critical decision, a never-ending process of weighing up dangers against opportunities, of balancing the possible presence of predators (and hostile humans) against the chance of new food resources and shelter. Thus, many specific aspects of nature which aided survival – trees which branch close to the ground, undulations in the landscape which offer view-points, the presence of large mammals – evoke an instinctive and favourable response in us still. The profounder implication of it all is that, in more general terms, there persists, deep inside us, deep in our genes, an immensely powerful, innate bond with the natural world.

The notion that we are part of nature, and nature is part of us, is of course not new; numerous pre-industrial societies, from Native Americans to Australian aborigines, have seen the world in this way (with their ways of imagining taken up by the modern Green movement), and many, many individuals have felt it, and often given it expression. But such notions of our unity with the biosphere have by no means entered mainstream thought, certainly among those people who administer the modern world, who make its decisions and run its governments and its corporations, and the countless millions who take their cue from them: rather, whatever their intrinsic value may be, such concepts have been largely ghettoised as anthropological or spiritual curios. The point about the idea of our bond with the natural world which comes out of evolutionary psychology – call it the bond of the fifty thousand generations, if you like – is that it is of a different order, for if it is true, as I believe it is, then it is not just spiritually true, it is also empirically true. It actually exists. It is a matter of fact.

But what might it mean to us? Powerful or not, might it be no more than a mere curiosity, a redundant evolutionary bequest that just happens to be there, like nipples on men? On the contrary, the bond seems increasingly to be of enormous practical importance for our physiological and psychological well-being, a phenomenon illustrated by the burgeoning research on the links between nature and human well-being, physical and mental. The study of this really took off in April 1984, when the prestigious journal *Science* carried a paper with one of those titles which, from time to time, make people around the world instantly sit up and take notice. It said: 'View through a window may influence recovery from surgery.' Its author, Roger Ulrich, an American architect who specialised in hospital design, had found that over nine years, patients in a hospital in Pennsylvania who underwent gall-bladder surgery made substantially better and quicker recoveries if they had a natural view from their beds. Some of the windows of the hospital wing looked out on to a group of trees, and some on to a brown brick wall, and those lucky enough to have the tree view, Ulrich found, recovered faster, spent less time in hospital, had better evaluations from nurses, required fewer painkillers, and experienced fewer post-operative complications than those who only had the wall to look at. Contact with nature, even if only visual, clearly had an empirical, measurable effect on people's physical and mental states; and since then, research on the practical health benefits of human involvement with the natural world has expanded enormously. A review of the literature to date published in 2005 reported that 'nature plays a vital role in human health and well-being', and suggested that contact with it should be an official part of all public health policies. The increasingly voluminous studies suggest that, even after five hundred generations, people are not really adapted to urban living and instinctively prefer natural environments to urban ones.

In fact, I would go further. I believe the bond is at the very heart of what it means to be human; that the natural world where we evolved is no mere neutral background, but at the deepest psychological level it remains our home, with all the intense emotional attachment which that implies – passionate feelings of belonging, of yearning, and of love. I said at the outset that the idea that we might *love* the natural world, as opposed to being aware merely of its dangers and opportunities, like the other creatures alongside which we have evolved, had long appeared to me an extraordinary phenomenon – but the bond of the fifty thousand generations is what makes it explicable. On the surface, in our everyday lives, this bond is largely invisible, it is very generally unfelt, as it has not only been overlain by the five hundred generations of culture which followed the conquering of nature by the farmers, but for those of us (since 2007 the majority of people in the world) who live in towns and cities in an increasingly hyperactive age, it is buried under an impenetrable mass of urban mental clutter. Yet deep down, it is there: we may have left the natural world, but the natural world has not left us.

And it can suddenly burst out. It can take you by surprise. You can sometimes not know quite what it is, why you are feeling as you are, why you are feeling so strongly:

> And I have felt
> A presence that disturbs me with the joy
> Of elevated thoughts; a sense sublime
> Of something far more deeply interfused,
> Whose dwelling is the light of setting suns,
> And the round ocean and the living air,
> And the blue sky, and in the mind of man;
> A motion and a spirit, that impels
> All thinking things, all objects of all thought,
> And rolls through all things.

You do not need to be Wordsworth looking down on Tintern Abbey to experience it; it is open to every one of us. Many people, faced with the beauty of nature, or its wonder, or its abundance, or the peace it can provide, or the sense we can feel in it every springtime of a world being reborn, have felt Wordsworth's joy; and I am no different. I have felt nature and joy many times since that first afternoon on the Dee, half a century ago, and as I have got older I have become increasingly convinced of something: that the joy itself, the intense love we can sometimes suddenly feel for the natural world, shows the existence of our continuing inner union with it better than anything else can.

That is why I am going to set out the ways in which I have met with joy over the years, just as you may have met with it in your life: I am going to present my evidence for the bond. I am not a scientist, not an evolutionary biologist nor a psychologist; I do not set out to prove it, to make formal argument, a step-by-step logical assembling of evidence. I am simply saying, this is what I experienced, and perhaps this will help towards understanding, for if our continued inner belonging to the natural world was not behind such intense emotion, what was? But it is indeed done in the hope that this novel feeling for what we are as humans, not yet forty years old, may subject to a wider awareness.

For the most important of the many important aspects of this new understanding is of course its context: this innovative sense of what nature really means to us, of what its value is, has come along at the very moment when we are tearing it to pieces. Just as *Earthrise*, the photo from space, showed us for the first time the planet's fragility and beauty, its uniqueness and isolation, so the insights of psychology and of evolutionary biology are showing us for the first time how we as humans are bound to it, bound to it in our souls inextricably, and how if we destroy it, we are destroying not only our home, which is dreadful enough, but also a fundamental part of ourselves which we cannot afford to lose.

And there, at last, is the possibility of a new defence of nature, one more robust and all-encompassing than either the hopeful idealism of sustainable development or the hard-faced calculation of ecosystem services; there, may be the beginnings of a belief and an argument with which to shield the natural world in the terrible century to come. The natural world is not separate from us, it is part of us. It is as much a part of us as our capacity for language; we are bonded to it still, however hard it may be to perceive the union in the tumult of modern urban life. Yet the union can be found, the union of ourselves and nature, in the joy which nature can spark and fire in us — even in the joy of the fifteen-year-old boy with his budget binoculars, listening to the cries of the wading birds borne upon the wind.

Terrible, though, is the word for it, the century that is coming for nature, and it is well under way. In fact, the destruction and the losses are already proceeding so rapidly and their scale is so colossal that a new problem arises: it is becoming difficult to describe them adequately, to do real justice to what each loss means, to expound them in other than the most generalised terms. You end up using statistics. I have done it here myself. *One in five vertebrates is threatened with extinction* . . . And perhaps it is worth considering that something vital can be missed, as the subject of environmental loss daily becomes more theoretical, abstract, and academic.

A prime example of this is the creation of two new metaphors to describe what is taking place. One is the Sixth Great Extinction. In the geological record researchers recognise five cataclysmic, life-extinguishing events in the earth's prehistory, beginning at the end of the Ordovician period 440 million years ago, when each time, the majority of species on the planet died out. Some of these events may have been caused by drastic changes in climate; others by the impacts of asteroids or comets, such as the object which hit what is now the Yucatan peninsula of Mexico at the end of the Cretaceous period, 65 million years ago, and wiped out the dinosaurs, in the most recent of the five. But such is the rate at which species are presently disappearing that many biologists consider we are today going through yet another major extinction, a sixth, which is comparable in scale to the rest – with the difference, of course, that this one has been caused by us.

The other metaphor is also inspired by the geological record, specifically by the idea of the geological timescale; it consists of a new (and so far, unofficial) label, the Anthropocene, to designate the epoch in which we are currently living. This is still formally considered to be the Holocene, from the Greek for 'wholly recent', covering the period since the end of the last glaciation in which agriculture began and civilisation took off; but so overwhelming has the human impact on the planet now become, above all on the atmosphere, whose composition we are so rapidly altering with such potentially disastrous results, that a growing number of scientists accept that the present time has a decisive character of its own and ought to be renamed as such. So welcome to the Anthropocene: the epoch when humans changed the planet.

They are very suggestive, these large-scale conceptions. Far-reaching images, such as the Anthropocene and the Sixth Great Extinction are, help us register the true degree of the planet's predicament and the real magnitude of the processes we have set in train which may bring about our ruin. They are of enormous value. They are talked about daily. Indeed, they are generating an academic industry on their own. But they do not necessarily convey the immediacy and astringent character of environmental loss, which in every case, somewhere along the line, involves hurt. If loss of nature becomes a sort of essay subject, we miss its immediacy; we may lose sight of its sadness and its nastiness, its sharp and bitter taste, the great wounding it really is. So before I take the high road to joy, I am going to return to loss, in a different way: not to the general or the broader picture, but to the specific. I am going to look at three particular examples of loss from my own experience, and the first follows on directly from my boyhood on the Dee.

We stand on the headland and look out over the plain, Nial Moores and I, the arid plain with a lorry trundling over it that till recently was a living estuary swept daily by the tides, an estuary holding flocks of wading birds so immense they were impossible to count precisely: twenty, fifty, seventy thousand, sometimes flocks of great knot, in particular, that may have been ninety thousand strong.

'In the telescope they were dark lines out on the flats,' says Nial. 'Then as the tide rushed in they would lift off, they would rise above the horizon and come in towards the shoreline in waves. Wave after wave. Great clouds of birds.'

I gaze at the extent of it, the estuary that is now dead, its flat surface of brown grassland like a vast empty dance floor stretching to that horizon, where it disappears in the mist. I find it hard to take in just how big it really is. Or was. I look to my left, I look to my right: in each direction it is all there is to see, before its distant vanishing. 'Yeah,' says Nial, 'makes the Dee look kinda cute, huh?'

There are a few cisticolas, small warbler-like songbirds, twittering in the grasses, and a solitary grey heron circling far away: that's it. All that remains of one of the earth's most remarkable concentrations of life, where the wader numbers reached four hundred or even five hundred thousand, where some of the most captivating creatures on the planet, curlews of different species, dunlin, great knot, bar-tailed godwits, grey plovers, Kentish plovers, Mongolian plovers, Nordmann's greenshanks, spoon-billed sandpipers, gathered in a profusion that was almost dreamlike. Now the monochrome monotony of the grass is broken only by scattered slabs of concrete and leftovers of rusty iron. The distant lorry trails a dust cloud. 'Know what you call this?' says Nial. 'A deadscape.'

Saemangeum in South Korea: the biggest destruction of an estuary that has ever taken place. It was a double estuary, in fact, of the Dongjin and Mangyeong rivers, in the province of

Jeollabuk-do, or North Jeolla, 160 miles south of Seoul, the capital, and it was 40,000 hectares in extent – three times the area of the Dee – with 29,000 hectares of this being tidal mudflats which at various times of the year hosted so many waders that it was by far the most important shorebird site in Korea ... perhaps in all of Asia. It was phenomenal. It was one of the wonders of the bird world. Now it is gone, snuffed out by a giant engineering vanity project, the building of the world's longest sea wall; a whole ecosystem annihilated. And standing here gazing upon it, at what it has become, and hearing at first hand exactly what it had been, I find welling up in me an unaccustomed emotion which I register with a shock as rage.

Nial Moores, British birder turned Asian environmentalist, discovered Saemangeum's true riches in 1998 when he carried out the first full assessment of the waterfowl and waders of South Korea's wetlands and coastline on behalf of a number of Korean environmental groups. It was genuine exploration, as military restrictions had previously made much of the coast inaccessible (South Korea being still officially at war with the North), and as he went round the country, sleeping in farmhouses, living on rice and seaweed and kimchee, Korea's signature condiment of pickled cabbage, feeling his way down unmapped tracks to the water's edge with a local activist and a taxi driver, he found nineteen sites that were of international importance for the numbers of shorebirds they held. In Saemangeum, he came upon El Dorado. 'It was quickly clear that the bird numbers were overwhelming. Eventually, we found a roost at the Okgu salt pans on the northern side of the estuary that was absolutely miraculous. The roost was of fifty to a hundred thousand birds. Just miraculous.'

But Saemangeum was already under threat. South Korea had decided in the 1980s to reclaim two-thirds of the tidal mudflats which fringe the coastline on its western side, and develop them for industry and agriculture; in 1991 it singled out the double estuary for the biggest reclamation project of all, to be facilitated by building, from its northern to its southern points, a sea wall more than twenty miles long which would cut it off from the tides and so choke the life out of it. The decision sparked a bitter fifteen-year battle between the South Korean government and the country's environmental activists. The environmentalists lost, and the resultant obliteration of this unparalleled habitat can be seen as one of the most egregious examples of environmental vandalism the modern world can offer. And yet, hard to credit though it may be, it is only part of a greater calamity still, the unfolding tragedy of the Yellow Sea.

The world has not yet woken up to this, but even in our terrible twenty-first century, it is likely to have few equivalents in terms of wildlife destruction. Peer at a map of east Asia and you will observe, enclosed by China on the left and the Korean peninsula on the right, what looks like a giant bay; and in effect, it is (albeit one 600 miles long by 400 miles wide), having once been a plain with a very gentle slope, which was covered by the rising sea levels at the end of the last ice age. It is called the Yellow Sea because it is coloured by the yellow-brown silt of the Yellow River, China's second longest, which carries a heavy sediment load; and the gentle slope of the shoreline and the silt load it receives combine with another factor, a very high tidal range, to give the Yellow Sea quite extraordinary wildlife value, which has only been recognised in recent years.

They mean that around much of its coastline there are tidal mudflats which are unusually extensive, and indeed may stretch for miles when the tide is out; and just as on the Dee, these black mudflats are the richest of all environments in terms of the invertebrates they harbour, the numberless congregations of molluscs and marine worms, of tiny crabs and crustaceans. For shorebirds, for waders, they have priceless, life-giving importance, and in fact the Yellow Sea tidal flats form the principal pit stop, the most important shorebird staging post, on one of the world's great migratory flyways.

The concept of the flyway is also fairly recent: it is a representation of the routes migrant birds use, shorebirds especially, on their annual journeys from the warm south in the winter to the insect-rich Arctic in the summer – 'from the tropics to the tundra' – and back again; and BirdLife International, the worldwide partnership of bird protection bodies, recognises and publishes maps of eight of them, eight mass transit systems, as it were, strung like thick vertical stripes around the globe. The Dee estuary, for example, is bang in the middle of the East Atlantic Flyway, the main route for the hosts of migrants which winter in sub-Saharan Africa and migrate in the spring up the Atlantic coast or over the Mediterranean, to breed in Europe and points north.

The Yellow Sea is also bang in the middle of a flyway, in this case the East Asia/Australasia Flyway, which may be something of a mouthful – it's easier to call it the EAAF – but which is, out there in the natural world, a wondrous phenomenon, older than history, as big as the weather, and something we are only now able to comprehend and visualise. It is the annual coming together of all the migrant shorebirds of the eastern half of Asia and all the migrant waders of Australia and New Zealand, in the stupendous springtime trek they make northwards to the tundra and coastline of Siberia to breed. Imagine it, looking at a map with China at the centre: two great streams, one from the bottom left, one from the bottom right, both pouring north and meeting halfway up, then flowing as a single stream to the top. Fifty million birds are thought to be involved.

The Yellow Sea is where the two streams join because it is the key staging point on the whole journey. Extensive areas of intertidal habitat, of mud exposed at low water where the birds can feed and replenish their energy levels, are actually quite rare, all around the world, and the Yellow Sea's concentration of them is essential both for the wader which has wintered in Burma and the wader which has wintered in New Zealand. The spring-time flight these birds make to nest in Siberia is more than five thousand miles, and it cannot be done without refuelling. The Yellow Sea tidal flats are where the refuelling takes place. They are the fulcrum around which the whole flyway is balanced. Fifty million wading birds, including some of the world's rarest species, depend utterly upon them. And they are being rapidly destroyed.

Reclamation is the process, and Saemangeum, which sits on the Korean side of the Yellow Sea, may be the most notorious example; yet what is happening cannot properly be understood without reference to China, whose Yellow Sea coastline is far more extensive. Modern China will have many impacts on the twenty-first century, but one of the most significant is the threat it poses to nature, which is ghastly. Nowhere on earth is the relentless process of wrecking the natural world being carried on more thoroughly than in the People's Republic, which, at the time of writing, seems poised to overtake the United States as the world's biggest economy (at least, by the measure of purchasing power parity). The phenomenal growth upsurge set off by Deng Xiaoping in 1978 has had two extraordinary results: it has pulled hundreds of millions of people out of poverty, and brought about the most concentrated burst of environmental destruction, desecration and pollution the world has ever witnessed. In somewhere like Shanghai, the wealth and the filth can be seen simultaneously. If you stand on the Bund, the elegant waterfront boulevard, and look across the Huangpu river to the skyscraper skyline of Pudong, the financial district, you behold a sight to compare with your first glimpse of Manhattan (it certainly took my breath away). Yet the Huangpu is not a waterway you would want to dip your toe in: in March 2013, for example, the local authorities recovered more than fourteen thousand dead pigs which had been dumped in it. Nor is the air of

Shanghai air you would always want to breathe: in December 2013 pollution hit such record levels that the cross-river panorama of Pudong could barely be seen in the smog, and parts of the city had to close down.

Nothing like China's current growth explosion has happened in history. It is hard to get your head around the scale of it: during the first quarter of this century, half of all the world's new buildings will be erected in China, and fifty thousand of them will be skyscrapers – the equivalent to ten New Yorks. It is similarly hard to take in fully the terrible environmental price that is being paid for this, which began to become visible to the outside world about a decade ago (a key moment being in 2007, when China surpassed the US in carbon dioxide emissions and thus became officially the world's biggest polluter). It is now increasingly documented, though, and you can pick out jaw-dropping figures at will: in 2006 the heavily industrialised provinces of Guangdong and Fujian discharged nearly 8.3 billion tonnes of sewage into the ocean, without treatment, a 60 per cent increase from 2001; by 2020 the volume of urban rubbish in China is expected to reach 400 million tonnes, equivalent to the figure for the entire world in 1997; and so on. But perhaps a single example should stand for all, that of the baiji, the legendary freshwater dolphin, a true wildlife treasure, 'the goddess of the Yangtze': by 2006, so gross and extreme had the industrialisation and pollution of the great river become that the baiji had been driven to extinction.

A major concern, however, is not only what China's frenzied growth is doing to its own environment, but what it is also doing to environments beyond its borders. The country is now not only the world's biggest importer of timber, it is also the biggest importer of illegally logged timber, and thus 'exporting deforestation'; its insatiable demand for wood is the major driver of rainforest destruction around the world. Its demand for ivory, especially since it was allowed to take part in the international ivory auction of 2008, is behind the renewed upsurge in the slaughter of African elephants; its demand for pangolins, or scaly anteaters, prized both for their meat and their scales, used in traditional Chinese medicine, has meant that all eight pangolin species are now threatened with extinction; its demand for tiger bones for medicinal use is similarly threatening the world's surviving wild tigers, while its demand for shark fins is behind the booming and unsustainable slaughter of the world's sharks (one estimate is that 73 million are killed annually for shark fin soup, increasingly served or ordered as a wealth-displaying dish by the burgeoning Chinese middle class). But the most far-reaching effect yet beyond its borders may be in what it is doing to the Yellow Sea, the vital flyway stopping place for the migrant shorebirds of twenty-two countries.

For China's wholly unchallengeable growth imperative, and the fact that six hundred million of its people, nearly a tenth of the world's population, live in river catchments which drain into the Yellow Sea, mean that the pressure to reclaim the tidal flats along its coastline is irresistible, and it is proceeding with ever-increasing rapidity. You can argue that such reclamation has always been done, but as a report by the International Union for the Conservation of Nature (IUCN) in 2012 makes clear, it is the speed and the scale at which it is being done now which is the problem. Since 1980 China has reclaimed no less than 51 per cent of all its coastal wetlands (this includes habitats such as mangroves and sea grass beds), and South Korea, 60 per cent (of a lower base). Of the key areas of tidal mudflats on which the shorebirds of the flyway depend, around the Yellow Sea as a whole, 35 per cent has already gone, and the remainder will go soon. It is likely that every major Yellow Sea tidal flat now has a development plan attached to it.

The situation, barely noticed by the world at large, is regarded by environmentalists involved as a wildlife catastrophe in the making, indeed, it is already happening, with the bird populations starting to fall: 'Observed rates of declines of waterbird species of five to nine per cent a year,' says the IUCN report, 'are among the highest of any ecological system on the planet.' Referring to the general effect on the flyway, the report says: 'The EAAF is likely to experience extinctions and associated collapses of essential and valuable ecological services in the near future.' The future of fifty million wading birds, and let it be said, of coastal fisheries on which thousands of people depend, is hanging by a thread.

There is perhaps one ray of hope for them, in terms of alerting the world before it is too late: the East Asia/Australasia Flyway has a poster species. The spoon-billed sandpiper is not only one of the most charming of all birds – the tiny wader has a spatulate bill, unique among sandpipers, which gives it a slightly comic air and the consequent endearing appeal of a puffin, while in its russet breeding plumage it is extremely pretty – it is also one of the rarest, and has long been at the top of the wish-to-see list of many birders (it certainly was with Nial Moores, and was one of the reasons for him coming to Asia). Breeding only in Chukotka, the Siberian province in Russia's far north-east, it winters five thousand miles away around the coastlines of Burma and Bangladesh, dependent like the other waders of the flyway on the Yellow Sea stop-over, and although it was never very common, by the millennium it was clear that it was declining catastrophically. In 2008, with the entire population now thought to be under two hundred pairs and falling at the rate of 26 per cent a year, it was listed as Critically Endangered.

Extinction seemed to be looming for 'the spoonie', but an international group of ornithologists decided on a desperate last throw to save it: they would establish a captive breeding population half the world away from its normal nesting sites, at Slimbridge in Gloucestershire, the wetland reserve founded by the painter and naturalist Sir Peter Scott, which is the leading centre of expertise in the conservation breeding of endangered waterbirds. And this was duly done. In 2011 eggs were taken from the wild in Chukotka and hatched locally in incubators, and the chicks were then successfully transported, not without taxing difficulties, to Gloucestershire. It was a wonderful wild-life story, one full of enterprise and hope and a certain amount of controversy (not everybody agrees with captive breeding), with at its heart as engaging and photogenic a creature as you could wish for, and I wrote about it, as did many other envir-onmental journalists: it was all over the BBC. As I researched the background, though, I started to see the larger picture and began to comprehend the apocalyptic threat hanging over the millions of migrant birds which depended on the Yellow Sea, with the rapid destruction of its tidal flats; and I came across this most extraordinary example of all of large-scale reclamation, this place called Saemangeum.

I'd never heard of it. The long battle to save it had passed me by entirely. But reading about it, I was gripped by the story of the giant sea wall which had done such harm to the natural world, especially after I looked it up on Google Maps and clicked on the satellite photo and suddenly there it was – the colossal barrage, seen from space, a bright white thin ribbon of concrete in the sea and behind it the huge dying estuary with its drying mudflats, no longer washed by the tides, no longer alive with shellfish and millions of other invertebrates, no longer peopled by numberless waders with their wild cries. It was astounding in its scale. It was unparalleled in the damage it had done. But more than any of that, it was everything I had always feared might happen to the Dee.

I know I was lucky. For a young person to fall in love with a place is surely one of life's greatest blessings, almost as remark-able a stroke of good fortune as being born into a happy family. It gives an electric charge to life and a sense of purpose which carries forward: existence henceforth cannot be seen as merely flat, boring, or meaningless. Whatever or wherever the place, whatever the nature of the love, it almost certainly involves an early acquaintance with beauty; an early acquaintance with worth; an early acquaintance with joy. And so it was with me at fifteen, with my parcel of wilderness.

For after I began to love it, in that sunlit October of 1962 when the world brushed up against Armageddon, the smiling old Pope threw open the worn-out shutters of the Church, and the Beatles began their dizzy ascent, I loved it dearly; I loved it as you might love a relative, an uncle, say, whom you had met for the first time in your teens, and who turned out to be quite exceptionally kind and intelligent and warm and wise: all at once there was something more, an unexpected and blessed presence in your life. I knocked a lot on that uncle's door, I went many times to the Dee and its untamed marshes throughout the rest of my adolescence, and I always went alone; my feeling was too personal to share. I think I almost saw it as a secret, the wildness of the estuary, although of course it was available to anyone; perhaps I mean, the nature of the feeling it produced in me, perhaps that was the secret, for you might expect to find that only in dramatic landscapes in distant lands, but I had stumbled upon it a mere bike ride from my suburban home.

I have tried to articulate it here with reference to the nineteenth-century Americans who understood so deeply why wilderness mattered, although I have to say in doing so, I have taken something of a liberty with timescales, for it was years before I read Thoreau and those who followed him; yet not long after my deep attachment to the Dee began, I did find my own expression of what the estuary meant to me in reading Gerard Manley Hopkins, the Victorian Jesuit whose poems, unpublished until thirty years after his death, had given some comfort to my mother in her distress, and who was himself torn between an overflowing joy in the natural world and a tortured guilt in the face of God (perhaps because of an unadmitted homosexuality).

I loved all his stuff, especially the famous poems like 'Spring and Fall' and 'The Windhover' and 'Pied Beauty', but one day I came across a lesser-known Hopkins quatrain which leapt off the page into my mind and has been present in the forefront of my consciousness ever since:

> What would the world be, once bereft
> Of wet and of wildness? Let them be left,
> O let them be left, wildness and wet;
> Long live the weeds and the wilderness yet.

There was my Dee, in ringing verse. Let them be left . . . I had no doubt then that they would be, the wet and the wildness. Just as the defining experiences of our youth, the ones which assume semi-mythical status in our own minds, stay with us all our lives, so it is natural to suppose that the places and circumstances which gave rise to them will similarly remain. Yet as the years passed such certainty faded, and I started to fear for the wetland that I loved. I started to fear that it might be destroyed.

There were two reasons for this. One was, that concrete proposals were put forward for its destruction. In 1971 a full-scale study was undertaken for a Dee Crossing and Reservoir Scheme, which would have turned it into two enormous artificial lakes behind a barrage topped with a new motorway from Merseyside into north Wales. It was a time when civil engineers were keenly attracted to the idea of throwing barrages across Britain's estuaries and reclaiming them: several of the biggest were in their sights, including Morecambe Bay, the Wash and the Severn, and over the following years, various versions of a Dee barrage were proposed, all of which would have meant an end to the wilderness and the untouched landscape, an end to the saltmarshes and the cries of the waders on the wind: an end to everything. Turned into reservoirs. Or farmland. Or housing. Or an industrial estate. I could easily see it happening. Why not? For the other reason I felt anxiety was a realisation that had begun to dawn on me: basically, nobody cared about estuaries. I had fallen in love with an anomaly. Most people saw the mouths of rivers as neither one thing nor another; they were the poor relations among landscape features, not remotely figuring in popular culture.

Don't you think?

I mean, know any estuary songs? Plenty about mountains and rivers and the sea, about forests and meadows and lakes. But estuaries? No. No one would ever speak up for them. No one would write their elegies when they were gone. They were regarded merely as between-areas, not here, not there, and their attractions were invisible; they were instinctively discounted.

So for much of my life I have had a nagging feeling that the landscape I fell in love with as a teenager was living on borrowed time, that somewhere so special could not really last, that sooner or later they would stick a barrage across its mouth like a giant gag and that would be that – telling my wife Jo, for example, when I first brought her to the Dee in 1991 and let her see it in panorama from the sandstone summit of Thurstaston Hill, that I was astounded I was still able to show it to her; and therefore, it is a cause of no small satisfaction to me that, partly because no barrage scheme ever did get off the ground, and partly because they did build the motorway to north Wales but they built it further south, and partly because the Royal Society for the Protection of Birds began buying up huge chunks of the estuary to defend it, and partly because formidable legal protection eventually arrived in the shape of the environmental laws of the European Union, the wilderness I stumbled upon all those years ago has not been destroyed, and every time I come round the corner on to the Parkgate promenade and see it stretching away, my heart lifts, as despite all the threats, despite all the plans for reservoirs and road crossings, despite all my fears, more than half a century on, the Dee is still there.

But Saemangeum is not. Saemangeum is gone. Extinguished. Rubbed out. The whole thing. It haunted me. I kept going back to Google Maps, spellbound by the satellite photo: so simple it seemed, the thin white line in the sea, stretching neatly from one point to another; such destruction it had done. There, but for the grace of God, went the Dee.

Yet no one seemed to be bothered about it.

It was over.

It was finished.

It was history.

It was only an estuary.

Who writes elegies for estuaries?

In the end, after thinking about it for three years, and seeking out some of the few people in Britain who knew about it and had actually been there, and hearing their stories, I resolved I would do it myself, I would go and bear witness to it, which is why, at the start of April 2014, I am standing with Nial Moores on the headland at Simpo, a small settlement in the centre of Saemangeum between its two now canalised rivers, the Dongjin and Mangyeong, looking out incredulously at the deadscape.

He is fifty, chunkily built. He runs Birds Korea, the conservation body he founded a decade ago, with a website in Korean and English. He speaks fluent Korean; his partner is a Korean woman; he seems to be almost morphing into a Korean himself, but not quite. He remains a British birder, a recognisable type (his brother Charlie is also a bird conservationist, back in Britain) with a passion that dates back to his early childhood in Southport, on the other side of Liverpool from my Dee, which he knows; among his first memories, from the age of five, are the calls of the wild geese, the pink-feet, flying out at night from the marshes behind Southport to roost on the Ribble estuary. ('I thought at first it was angels' trumpets.') He came to Korea in 1998 by way of Japan, where he had spent eight years, shifting from his teaching career to become a full-time conservationist; he learned

Japanese and became deeply involved in environmentalists' struggles to save threatened Japanese wetlands. Eventually, Korean environmentalists invited him over to share his birding expertise in the survey he carried out for them, which disclosed the full wonder of Saemangeum; and he stayed.

He tells me the history of the fight to save the estuary, a determined and bitter battle, involving lengthy court cases and many demonstrations, most movingly, the *samboilbae* of 2003. The word means 'three steps and a bow' and refers to a procession thus performed, with those taking part taking three steps, dropping to their knees and bowing down to the ground – touching it with head and elbows – then getting back up and repeating the process. It is enormously demanding, spiritually as well as physically, but to express their sympathy with the creatures that would die in the estuary's destruction, in the spring of 2003 two Korean Buddhist monks and two Korean Christian ministers led a samboilbae all the way from Saemangeum to Seoul. It took them sixty-five days to complete, in all weathers, and eight thousand people met them in the capital. But even that was not enough. In April 2006 the final gap in the sea wall was closed, and the estuary's fate was sealed. There were demonstrators at the barrage that day, but Nial went birding elsewhere: he could not bear to watch it.

He is torn between his love for his adopted country and what he sees it doing to the natural world. It has already obliterated three-quarters of its tidal flats, he says, and he thinks reclamation is being carried out now for reclamation's sake. 'It's so sad. I feel love for Korea. I want to be part of Korea. But this . . . it's just so disastrous. It's on an unimaginably huge scale. It's hard to get my head around it even now, and I look at it hundreds of times.' I was in agreement with him. My impressions of South Korea were unhappy ones. I liked the Koreans I met and I enjoyed the different, fiery food, but I beheld a country rapidly destroying its own beauty by its economic growth mania: it was China writ small. They had performed a similar trick, they had pulled their people out of poverty, from a national income of less than $100 a head in 1960, on a par with some countries of sub-Saharan Africa, to $33,000 a head fifty years later, when it had become the twelfth largest economy in the world; but the environmental price, as with China, was appallingly high. The frenzy of construction struck me most forcibly. The country seemed obsessed with building things, with piling up more and more infrastructure; it had to be constructing new roads everywhere even though an extensive national highway system had been in place for years and it didn't seem to need them; it had to be not only putting up new bridges and dams and industrial complexes and ports left, right and centre, and block after block of office buildings, but to be tearing down what was there already and rebuilding it, whatever it was. There were very, very few old buildings; most if not all of the ones I saw, even the tourist attractions, turned out to be replicas: new old buildings. Historic villages, ten years old. 'If they have a nice river, with meadows by it, where people like to walk,' said an English acquaintance in Seoul, the capital, 'They can't just leave it alone. They'll develop it, and turn it into an eco-park. That's the Korean way.' I felt the construction mania had reached the stage where it was blighting the land, which is not vast and cannot absorb an unlimited pummelling without showing the bruises: I only spent a week there and I only travelled a few hundred miles, so of course I missed huge amounts of the country, but I did not come across a single landscape in my time in South Korea that I would describe as unspoiled.

Saemangeum exemplified it all – that is, what you get sooner or later if you are wholly consumed by the obsession with economic growth: the deadscape. We looked at it from several places and from several angles and it never ceased to amaze, the scale of it, and the scale of what had been lost, recounted to me vividly by Nial, and also the fact that the estuary might have been reclaimed, but eight years on, the reclaimed forty thousand hectares had still not been put to use. It was just an empty brown plain of rough grass. No industry, no agriculture, no housing. Nothing whatsoever. Why had this colossal project been so essential, if eight years after completion, the authorities were still wondering what to do with it? It seemed more than anything to be development for development's sake, a view which was strongly reinforced when we went to see the ostentatious artefact that had caused all the trouble.

There was no doubt that it was very big, the Saemangeum Sea Wall. Bigness, you would have to say, was its basic, unignorable characteristic. It took half an hour to drive from one end of it to the other, non-stop, with the Yellow Sea on one side and the extraordinary estuary which it had extinguished, now the deadscape, on the other; when you started out along it, you could not see the finish. But the more I saw of it, the more I felt it had another characteristic I had never before in my life associated with a piece of civil engineering: there was something false about it.

For a start, it was boastful: it not only stretched for 33 kilometres, so yes, it can be seen from space, but it had purposely been constructed 500 metres longer than the Afsluitdijk, the barrage which turned the tidal Zuiderzee in the Netherlands into the freshwater Ijsselmeer, with the specific object of replacing this latter in *The Guinness Book of Records* as the longest sea wall in the world. (There is doubtless a bragging press release somewhere with the figure of just how many million tonnes of concrete went into its construction, although it has passed me by.) It was as if the purpose of the whole development was not need, but vanity. Look what we can do! Biggest on the planet! Yet there was something worse than that, something repellent: the whole thing was dressed up, and it was dressed up in lies. All the signage along the sea wall strove to create a sort of ersatz enthusiasm, a sort of gimcrack cheerfulness, beginning with the road signs themselves, which were a bizarre combination of stern safety exhortations with what was meant to be homespun Asian folk wisdom:

No Stopping On The Road

Fishing Is Forbidden

Be Happy For A Hundred Years

They were complemented by sizable roadside advertising hoardings with slogans which were pitiful in their speciousness, such as:

Saemangeum – Land Of Hope

or

Saemangeum – Dream Of The Future

or

I ♥ Saemangeum

Most grating of all were the frequent attempts to portray the project as green, in the environmental sense of the word, ranging from labelling the concrete parking places Dolphin Bay or Sunset Bay to publicity posters illustrating smiling, attractive young families looking admiringly on plans for Green Saemangeum, which displayed housing schemes with watercourses and flying creatures vaguely resembling wading birds, of no recognisable species.

This, to portray the project which had done more damage to shorebird habitat than any comparable project in history; this, to gild the image of the narcissistic construction which had wiped out an estuary that was wondrous, for no discernible purpose.

I thought the whole thing had been put together by people who had a substantial part of their moral compass missing.

It was public relations at its most pathetic.

It was nauseating.

Nial wanted me to get a feel for what the birdlife had been like in the estuary before the sea wall snuffed it out; and so, if I write Saemangeum's elegy here, it is not without having some personal sense of what has been lost.

He took me to the next estuary going northwards, that of the Geum river. The tidal flats on its south side have been consumed by the expanding port of Gunsan, but on the north side, which is in a different administrative distict, Seocheon county, the remainder had up till then been protected by the environmentally minded Seocheon mayor, Na So-Yeol, although schemes were constantly being put forward for the remaining flats to be reclaimed; and on the drive there, we came across a half-built road which was pointing directly towards the estuary, as if its promoters were just waiting for the moment when they could finish the job and share in the development bonanza which another tidal flat reclamation would provide.

Nial took me to watch the high tide come into Janggu bay, on the estuary's outer edge; there was a tiny concrete quay where inshore fishing boats brought in substantial cargoes of what seemed to be cockles, which made a perfect observation platform for the bay as a whole. Almost the first bird we saw was a far-eastern curlew, which is the world's biggest wader, its decurved bill being even longer than that of its cousin the

Eurasian curlew, the curlew of the Dee with its melancholy bubbling song – numbers of which were also in the bay.

We watched the far eastern curlew catch a crab from the black mud and with eye-catching deftness, snap off its legs on one side, turn it around in its bill, and then snap off the legs on the other, before swallowing it. The bird, Nial said, bred in Arctic Russia and wintered in Australia. It had a global population of just 41,000, which, in the IUCN categorisation of threatened species, made it Vulnerable; there were eight species breeding or refuelling on the South Korean tidal flats which were now threatened, he said, with Saunders's gull and the relict gull, the Chinese egret and the great knot also being classed as Vulnerable; the black-faced spoonbill and Nordmann's greenshank being classed as Endangered; and the spoon-billed sandpiper, Critically Endangered. All depended absolutely on the Yellow Sea flats which were being so speedily destroyed, he said. 'Here is a group of species really heading towards rapid extinction, and no one's talking about it.'

As the sea crept in over the outer flats, wader flocks began to drift into the bay in a steady stream, shimmering clouds of birds settling on the high tide line; Nial, who was expert in counting them, ultimately reckoned there were more than 13,000, including 500 grey plovers, 2,000 great knot, 2,500 far eastern curlews, 3,000 dunlin and more than 5,000 bar-tailed godwits, these last having just arrived from Australia and New Zealand, en route for Siberia. The waders with their wild allure, and their paradox – the gift to us of mud. The mud which will not long remain now. Thinking of the mortal threat to them, I took delight in their abundance, even if the numbers were but a fraction of those which Saemangeum, the living estuary, the wonder of the shorebird world, had once hosted. I mourn Saemangeum's passing; may its memory remain.

We watched the birds in their nervy congregations, skittish, flickering in the sunshine, for an hour or more; then the tide

retreated, and the flocks began to take off in colossal whisper-
ings to return to their feeding grounds on the outer estuary;
and as they did so, I became aware of a plangent metallic noise.
It was the rhythmic clanging of a hydraulic hammer, a pile-
driver banging in the piles for some major piece of construction
somewhere in the vicinity, perhaps even the road, the half-
completed road which pointed towards the Geum estuary and
its tidal flats, waiting for development to be given the green
light. And then I became aware of another noise: a Eurasian
curlew was simultaneously bubbling its melancholy spring song,
the song I learned to love half a century ago, on the Dee:

> Ho, hullaballoing clan
> Agape, with woe
> In your beaks . . .

The two noises combined in my ears and seemed to coalesce
into the whole tragedy of Saemangeum, of the fact that even
the natural world's most remarkable sources of life are now
being steamrollered by development, by the frenzied and unchallengeable rush for growth, by the monstrous uncontrollable
runaway scale of the human enterprise: here was the curlew,
bubbling the spring, and accompanying it the hammer, clanging
and tolling its doom.

4

The Great Thinning

So South Korea has destroyed Saemangeum, the bay with its wealth of shorebirds almost beyond imagining, and China has destroyed the baiji, its dolphin, its treasure, the goddess of the Yangtze, in its remorseless fouling and desecration of so much of the natural world . . . but my own country of Britain is no better. In my lifetime, it has wiped out half its wildlife.

This is a quite remarkable historical event which, although it is now at last understood by experts in conservation, has by no means fully penetrated the public consciousness. People may think of how Britain has changed in the last half-century as a country which has lost an empire but become in the meantime more wealthy, more multicultural, more tolerant, and less class-bound – yet hardly anyone would instinctively think of it as a country which, in that same relatively short time, has annihilated half its biodiversity. It is too recent a realisation. Asked to dash off a portrait of the modern United Kingdom for a sophisticated audience, a writer-at-large from the *New York Times* or *Le Monde* or *Corriere della Sera* would never in a million years home in on this change, which to me has altered the character of my native land as profoundly as has immigration, or the end of social deference, or the coming of sexual equality.

In wildlife terms, the country I was born into possessed something wonderful it absolutely possesses no longer: natural abundance. There was a profusion of life forms all about us, from house flies to house sparrows, even in the suburbs: there was a richness of weeds, there was a richness of pests, there was a richness of organisms that were glorious – the August buddleia of suburban Sunny Bank was covered in glories – and in the countryside, of course, all this was enhanced a thousandfold. Abundance gave exhilarating substance to life, to normal life, which we took entirely for granted: it seemed to be the natural order of things. Indeed, it had been there for untold centuries. It was one of the reasons I fell in love with nature, when I first sauntered out into the countryside of the Wirral as a skinny kid in the fifties to collect birds' eggs and catch butterflies and fill glass jars with newts (to live in a washing-up bowl in our garden shed until they died, God forgive me) – and the Wirral possessed nothing of the wildlife opulence of somewhere like Dorset. But it was rich enough, and nature so rich was easy to love. Now the riches have gone, and the wildlife that surrounds us in our everyday lives in Britain, and beyond us in the greater part of the rural landscape, is, with a few exceptions, impover- ished, scanty, and sparse; wildlife that is worthwhile can still be found, but it must be sought out. Abundance, blessed, unregarded abundance, has been destroyed.

That this should happen, that a country should lose half its wildlife in little more than half a century seems unimaginable, scarcely to be believed – is there a historical parallel? – but the figures are there, in the most comprehensive datasets in the world for the changing status of at least three major groups of species: birds, wild flowers, and butterflies. Their British pop- ulations have all been devastated. The key period of loss was probably between 1960 and 1990 (although it began before and has continued strongly since), yet the scientific recognition of the true scale of the losses, taken in the round, has only come about since the start of the new millennium. Some of my generation, the baby boomers, felt it in their bones, some of them sensed things were changing profoundly, but mostly their lives were too full, privileged, and enjoyable to stop and look closely, and anyway, they were approaching retirement before the full astonishing picture emerged of what had been destroyed.

The engine of this destruction took British society by surprise: it was farming. One of the key characteristics of wildlife in Britain, especially in the lowlands of England, is often overlooked because it is so obvious: it exists on farmland. It has nowhere else to go. This is by no means the case in tropical countries, say, or even in the United States, where you would not think of taking a wildlife holiday in somewhere like the grain prairies of Kansas; you would go to a wilderness area such as Yellowstone, the national park. America is so big that it can happily have separate locations for large-scale agriculture and for wild things. But Britain cannot. It is small and its countryside has long been an intimate mix of habitats where wildlife and farming have had to coexist and traditionally did so; indeed, this is what gave the scenery its celebrated loveliness and charm. A cornfield did not just contain corn, it contained blood-red poppies and glowing blue cornflowers as well, and clouded yellow butterflies flew about it and skylarks sang above. It was a landscape that delighted.

In my boyhood, people worried about threats to the countryside, but the concerns focused on development, on the siting of new factories and new towns, on piecemeal 'ribbon development' of new housing along rural roads, and especially on such excrescences as American-style advertising hoardings, or the march of tall electricity pylons across cherished views. Thus in 1947 the Town and Country Planning Act set up a legal system specifically to keep all this in check, to make sure that the actions of individual persons or companies, in development terms, were in line with the wishes of society as a whole. No one, no one at all, foresaw that agriculture itself would be the wrecker. Farmers were respected and seen in the public mind as the eternal guardians of the countryside and its wildlife, and consequently excluded from the planning system and in no way bound by its constraints. As Mr Schwarzenegger would say: *Big Mistake*.

For after the Second World War two major changes came to farming in Britain. The first was new technology: immensely powerful new agricultural machines, chemicals and techniques. The second, even more important, was the economic pressure to use all this to the uttermost, to squeeze every last penny of profit from the land. The whole process is referred to as intensification, and the shift to intensive farming was initiated by the government itself, by the post-war Attlee administration, which was obsessed by how close the German U-boats had come to cutting off Britain's imported food supplies during the war. Britain needed to be self-sufficient in food, they felt, and production needed to be expanded drastically. So British farmers were for the first time given a guaranteed price for whatever produce they were minded to grow, and if their guaranteed minimum price fell below the market price, then they would get payments to make up the difference.

This price support meant that for the first time it became profitable to plough up any marginal land, any piece of downland or moor or scrub or damp pasture which had not previously grown crops, but which, conversely, might be very wildlife-rich. Furthermore, generous capital grants were made available to do this, and to get rid of any pesky obstructions in the way of the big new modern machines, such as hedges, copses, woods, ponds, and ditches – landscape features where wildlife, of course, had also long flourished. A frenzy of subsidised bulldozing began, of hedgerows especially, thousands of miles of them disappearing, some of them hundreds of years old, as the landscape, especially in eastern England, was reshaped from the traditional patchwork-quilt of small to medium-sized fields variegated with their hedges and woods and quirky corners, to vast plain prairies of wheat and barley, Kansas-style. And let's not forget the orchards. Age-old orchards, atmospheric groves of gnarled, lichen-covered trees which had borne fruit time out of mind. They were grubbed out by the hundred.

All of this, energetically encouraged by the Ministry of Agriculture, naturally began to affect wildlife, and the damage was compounded by two new farming techniques, still barely comprehended by the public at large, the change from spring-sown to autumn-sown crops, and the shift from hay to silage. New varieties of crops sown in autumn were known to be more productive and could be harvested much earlier than traditional spring-sown cereals, in July or August rather than September, but they had enormous disadvantages for wildlife, particularly birds, which were doubly hit. Autumn sowing meant that the stubble from the previous harvest, rich in seeds which flocks of songbirds such as finches fed on right through the autumn and winter, was now quickly ploughed back into the ground, and the finches' food supply disappeared; and when the spring came, the autumn-sown crops were already up in the fields, and so high that other farmland birds such as skylarks and lapwings could no longer nest in them.

The disappearance of haymaking, one of the age-old activities in the farming calendar, and with it of hay meadows and their replacement by fields of artificially fertilised ryegrass, was an even more detrimental change. Hay is grass which is cut just the once, in late summer, then dried so it does not rot when kept in a barn, and fed as fodder to horses. In 1950 there were still three hundred thousand horses working on British farms, but thirty years later they had nearly all gone, outpaced by the march of the machines. With them went the need for hay. Cattle could be fed on something else: silage, grass cut while still green, which rots down into a gunk that cows find perfectly accept-able. A special variety, perennial ryegrass, was discovered to be ideal for the job, and to make it grow as quickly as possible it was treated with large amounts of artificial fertiliser, meaning it could be cut at the beginning of June or earlier, and then cut again six weeks later, and perhaps for a third or even a fourth time before summer was over.

All across England, ryegrass fields began replacing the hay meadows and ancient grazing pastures which had been among the countryside's great delights, as they were botanical treasure houses crammed with wild flowers such as buttercups, red clover, yellow rattle, lady's bedstraw, knapweed, green-winged orchid, ox-eye daisies, kidney vetch, and many many more, presenting an animated chaos of colour which could dazzle the eye. Abundance at its most enchanting. It is thought that about 97 per cent of them have gone now. In the ryegrass leys which replaced them, there was but one species only: ryegrass. It was green concrete, as the phrase has it. It had been so heavily fertilised that it out-competed any other plant; nothing else could survive. The process was known as 'improvement'. (Traditional grassland, where it survives, is now referred to as 'unimproved'.) But improvement was damaging to more than wild flowers. The various species of birds which nested or foraged in traditional meadows, such as corn buntings, quail, whinchats, and (once upon a time) corncrakes, had in the past always had room to breed successfully before the hay was cut in late July. But with the much earlier cuts for silage, usually repeated, their nests and eggs and chicks were mangled by the machines. Continually. They were as doomed as the sparkling wild flowers were.

That might all seem bad enough, the bulldozed hedges, filled-in ponds, grubbed-out orchards, the destruction of the autumn stubbles, the destruction of the hay meadows; but Farmer Giles, the ruddy-faced commonsensical custodian of the countryside, who knew so much better than those folk in towns about how to look after it properly with his age-old wisdom, or so the poor deluded population of the British Isles continued to think, put the tin hat on it all with poisons. Agricultural poisons. Poisons for this, poisons for that. Kill off the insects. Kill off the snails. Kill off the wild flowers. Kill off anything that isn't your money-making crop with herbicides, pesticides, fungicides, molluscicides . . . Farmer Giles loved them all, he turned on the tap and let a great flood of poison wash over the land, which, God help us, floods over it to this day. He began just after the war with the chlorinated hydrocarbon, DDT, the first of a new generation of synthetic pesticides – compounds made in the laboratory rather than from naturally occurring substances – and quickly followed it by spraying his fields with the even more powerful organochlorines, such as aldrin and dieldrin. The problem was, these things didn't just kill insects; it turned out they killed birds as well, in startling numbers (and for good measure they quickly killed off all the otters of lowland England, probably because they built up in the fat of eels, a favourite otter prey item). It was several years before anyone noticed that English otters seemed to have vanished, but the corpses of thousands of dead birds were all too visible at once, and in America especially they provoked a fierce public outcry, led by Rachel Carson with her magisterial exposé and denunciation of the agrochemical industry and all its works, *Silent Spring*, whose publication in 1962 can be taken as the start of the modern environment movement.

Dead robins littered across suburban lawns could not be ignored, no matter how loudly the US chemical industry screamed that Rachel Carson was an hysterical woman (the adjective and the noun each being half of the accusation), and so eventually DDT, aldrin, dieldrin, and other organochlorine compounds were banned, in America and Britain both. They were replaced with new generations of pesticides which generally did not kill birds directly, such as carbamates, organophosphates, pyrethroids, and more recently, neonicotinoids. I say generally, although some of these compounds were still bird-toxic, such as carbofuran, a carbamate which (although banned) remains the substance of choice for British gamekeepers who want to poison birds of prey, or parathion, an organophosphate used to kill bird pests such as queleas in Africa. Certainly, however, they all killed insects, and they did not just kill 'target' insects, they killed almost all insects, just as herbicides usually killed almost all herbs, the vast majority of the flowers which had added to the beauty of cornfields (and in killing the insects and the non-crop plants they killed off the food supply for the birds); and in due course their widespread use became routine. For Farmer Giles, second-nature, even. He couldn't get enough of them. This is the heart of the matter. The incorporation of the wide-spread use of deadly poison into everyday agricultural practice is what, above all else, has destroyed the wildlife abundance of my country and I curse it. You can say it's essential for the produc-tion of food. I say, no it isn't, not on the scale in which it has been used. It has dealt a mortal blow to half the life of the land in which I grew up.

That all this was happening took a very long time to dawn on the public; in fact, it was more than three decades before people began to perceive properly how intensive farming was wrecking the natural world, so strong was the Farmer Giles custodian myth. In the intervening years many rural communi-ties had seen long-beloved pieces of countryside swept away, despoiled or changed beyond recognition by farmers chasing ever more profits, and had had their protests or passionate appeals to desist contemptuously refused, with no one in officialdom to turn to: farmers were outside the planning system, remember, and could do whatever they liked with their land – just as Genesis 1:28 had urged them to.

It was not until 1980 that the anger at what they were doing

finally exploded, through a groundbreaking book by the environmental campaigner and academic Marion Shoard. *The Theft
of the Countryside* set out for the first time, in detail and at
length, the results of British farmers' remorseless pursuit of
money at the expense of the traditional and dearly loved rural
landscape, a pursuit which, although already calamitous, had
been given a savage upward twist by Britain's entry into the
European Economic Community (as the European Union was
then called) in 1973. The EEC's bizarre Common Agricultural
Policy, which the French wanted in order to keep their thousands of small farmers on the land, and the Germans went along
with because they wanted to get back into the human race after
the Second World War, abolished the laws of supply and demand.
Instead, everyone was to produce as much food as they possibly
could, and even if there was no demand for it, the price would
not fall! Consumers didn't have to buy it – the Community
would buy it! Then stick it all in warehouses, millions of unwanted
tons of it – a butter mountain, a wine lake, but never mind,
just keep producing, rip out every hedge, fill in every pond, kill
off every insect and wild flower that might possibly be competing
with you, ruthlessly turn all your acres into a clinical food
factory and Brussels will definitely make it worth your while.

Marion Shoard shouted at Farmer Giles and his colleagues
the first properly heard *I accuse!* She explained to a hitherto-
ignorant public how if a farmer wanted a grant to destroy, in
effect, a piece of well-loved landscape, there was no test of
appropriateness, no means test, and it made no difference whether
the commodity he was going to produce was in surplus or not:
he was just handed the money, and there was no limit on the
number of grants that could be given in any year. No official
environmental oversight whatsoever took place of agricultural
intensification, such as the planning system provided for development – the Ministry of Agriculture cared not a jot about
that, it was simply not in its DNA, all it cared about was

maximising production – and there were virtually no restraints at all on what farmers could do. Shoard instanced case after case where captivating countryside, such as Graffham Down in Sussex, had been turned into cereal desert despite the ardent and heartfelt objections of local people. If a picknicker dug up a cowslip root, she pointed out, he or she could, under the Conservation of Wild Creatures and Wild Plants Act, 1975, be prosecuted; but if a farmer ploughed up a whole field of cowslips, no one could do anything about it.

The Theft of the Countryside was impassioned polemic as well as documentary and it shot a timely hole in the myth of the farmer as countryside guardian, and was widely noticed: the sculptor Henry Moore wrote the foreword. But although it was the first major stocktaking of the damage intensive farming had done to Britain, it was essentially about landscape rather than wildlife: of the book's 272 pages, only eight were given over to 'Disappearing Wildlife' specifically. That was understandable. Ravaged landscapes were readily visible, but the decline of wild-life populations, as it was happening, was then much harder to register. For the losses were of numbers, not of species. Consider: the principal metric the general public instinctively uses for wildlife loss is species extinction. National (and of course, global) extinctions of species are always observed and remarked upon. If they happen, everyone knows something's wrong. But in Britain, national extinctions were not piling up like ravaged landscapes were. What was happening with wildlife was more subtle: it was a great thinning-out of all populations.

Right across the land, the once unsullied land now bulldozer-battered and awash with poisons, there was simply *less* of every-thing, year on year: fewer birds, fewer wild flowers, fewer butterflies. Species A might still be around; there just weren't so many examples of it; and the process, of course, was cumu-lative. My experience of having written about this as a journalist, and having had extensive reader response, is that not a few people sensed it intuitively, but could not quite put their finger on it. Was it really happening? Were they imagining it? They were uneasy, but not certain. It is only now, another thirty-five years on, that we are certain, that we know it was happening indeed, and this is because of a fortunate development which began in the 1960s: the setting up of a series of long-term wildlife recording schemes by Britain's extensive community of naturalists, professional and amateur. These schemes, first to record wild flowers, then birds, then butterflies, were able to track precisely Farmer Giles' pitiless onslaught on the biodiversity supposedly in his care, although they began when the process was well under way, so the baselines used were already degraded ones and did not reflect the position of, say, 1947; thus, the true losses are substantially underestimated by figures we have now. Nevertheless, they still present an irrefutable picture of staggering decline, not by national extinctions – which would have made headlines, and long ago alerted the nation to what was happening – but in general abundance.

With birds, for example, there were only two national extinctions in Britain in the post-war period, those of the red-backed shrike and the wryneck, both charismatic species, alas (although both have returned to breed intermittently). But the number of birds which have declined so much as to be *locally* extinct, over great swathes of the land, is hugely higher. In the period 1967–2011, according to the Common Bird Census (and its successor scheme, the Breeding Bird Survey), the turtle dove declined in Britain by 95 per cent, the grey partridge by 91 per cent, the spotted flycatcher by 89 per cent, the corn bunting by 88 per cent, and the yellow wagtail by 73 per cent, while the tree sparrow declined in England alone by 95 per cent; and so it goes on. In most of the country, they've simply vanished. The position is precisely paralleled with wild flowers. During the whole of the twentieth century, there were about twelve national extinctions among Britain's fifteen hundred or so native plants (the figure varies with how you define native), including such colourfully named species as thorow-wax, swine's succory, narrow-leaved cudweed and summer lady's tresses; and you might think, all things considered, that was not too bad.

But in the millennium year the naturalist Peter Marren, on behalf of the charity Plantlife, looked at the position on a county-by-county basis using the county *Floras*, those comprehensive wild flower catalogues which are such a feature of British botany – *The Flora of Cambridgeshire, The Flora of Kent* – and a dramatic and alarming picture emerged when it was local rather than national extinctions which were examined. Over the century Britain as a whole might have lost its dozen or so species, but Marren calculated that Northamptonshire, between 1930 and 1995, had lost 93; Gloucestershire, between 1900 and 1986, had lost 78; Lincolnshire, between 1900 and 1985, had lost 77; Middlesex, between 1900 and 1990, had lost 76; Durham, between 1900 and 1988, had lost 68; Cambridgeshire, between 1900 and 1990, had lost 66; and so on. (These figures were later re-examined and somewhat revised down by the botanist Kevin Walker, but even the revised figures are astounding.) And it is exactly the same story with butterflies. There have been only three national extinctions in the post-war years, those of the large tortoiseshell, of the large copper – which had been reintroduced, having gone extinct once before – and of the large blue (which has now been reintroduced itself, with great success); but since the butterfly recording schemes first started, nearly three-quarters of our fifty-eight remaining species have declined and disappeared over much of the country. Between 1970 and 2006, for example, the high brown fritillary declined in distribution by 79 per cent, the wood white by 65 per cent, the pearl-bordered fritillary by 61 per cent, the white-letter hairstreak by 53 per cent, and the Duke of Burgundy by 52 per cent. As some of these species were uncommon anyway and the declines were from what was already a very low base, they were left in a parlous position: the high brown fritillary is now critically endangered in Britain, and the other four species just mentioned are all regarded as endangered, as are several others.

All across the land, they tumbled in numbers, the birds, the wild flowers, the butterflies, and it is clear that more than half of all Britain's wildlife, as it existed at the end of the Second World War, has now gone. If we were to take just a single statistic, the key one would be the farmland bird index figure from 1970 onwards, which the government now publishes. The most recent index, for 2013, showed the combined population of nineteen farmland bird species, from skylarks to lapwings, from grey partridges to yellowhammers, at 56 per cent below the 1970 level, so even by the British government's own admission, we have lost, just since The Beatles broke up, more than half the birds which once so delighted visitors to the countryside, and as they had been declining for virtually two decades before the index start date, the real figure is obviously much larger; and so with the insects; and so with the flowers.

This was the great thinning; this was the destruction of wild-life abundance in my country, at the hands of the farmers. As I was born in the same year as the piece of legislation which first gave them their price support and thus launched intensi-fication – the Agriculture Act, 1947 – the process has precisely paralleled my own life. I was just lucky that the natural plenty of wild things survived long enough for me to encounter it in childhood and for it to leave an impression that has never faded, even though the profusion itself has melted away entirely and a child born in the generations after mine can never know it. It's bad enough that individual species are hard to find. The pearl-bordered fritillary, for example, a jewel of an insect, was described thus in J. W. Tutt's *British Butterflies* of 1896 ('A Handbook for Students and Collectors'): 'This is a very common woodland butterfly in England, haunting the flowery openings and sides of almost every wood of any size.' You will travel a long way to see one now. Try finding a corn bunting, stout yet perky with its song like a bunch of dropped keys, or even more, try finding any of the cornfield wild flowers, the arable plants, which once splashed the crops with colour – cornflower, corn marigold, corn buttercup, pheasant's eye – they survive in just a few secluded corners. But even more than the single species, it's the loss of abundance itself I mourn, and I know that some of the baby boomers, in whose lifetimes it disappeared, mourn it too. People over the age of fifty can remember springtime lapwings crying and swooping over every field, corn buntings alert on each hedge and telegraph wire, swallow aerobatics in every farmyard and clouds of finches on the autumn stubbles; they remember nettle beds swarming with small tortoiseshell and peacock caterpillars, the sparkling pointillist palette of the hay meadows, ditches crawling and croaking with frogs and toads and even in the suburbs, songbird-speckled lawns and congregations of house martins in their dashing navy-blue elegance . . . but most vividly of all, some of them remember the moth snowstorm.

Moths have long been unloved. There are about a dozen mentions of moths in the Bible and all of them are unfavourable: they are wretched little brown things akin to rust, which eat your clothes, as well as your books and your tapestries, if you believe the Good Book, and nothing more. The prejudice has been persistent: people have for centuries seen moths as haunting the night, like owls and bats, like ghosts and goblins and evil-doers, and thus sinister and shudder-provoking, whereas butterflies, their relatives, eternally symbolise sunshine and have been adored. Yet in my own country of Britain, perceptions are changing. Lovers of the natural world are becoming more and more drawn to moths, many of which are every bit as big and as bold in their colour schemes as butterflies are, such as the black and cream and orange Jersey tiger, or the pink and green elephant hawkmoth, or even the legendary Clifden nonpareil, the outsize and shadowy species which shows on its underwings a sumptuous colour found nowhere else in the moth world: lilac blue. The difficulty of seeing them at night can easily be circumvented with a moth trap, essentially just a powerful light attached to a box, which exists in several designs but is always based on the same principle: moths are attracted to light, moths fall into box, moths settle down and go to sleep, and then can be released perfectly unharmed in the morning – after you've had a close look at them and identified them. This may seem like prime nerd territory, and sure, it may be, but the number of nerds is soaring: according to the charity Butterfly Conservation, there may now be as many as ten thousand enthusiasts in Britain operating moth traps in their gardens on summer nights. I am one of them.

When you do that, you start to realise for the first time a basic wildlife truth: it is moths, not butterflies, which are the senior partners in the order Lepidoptera, the scale-wing insects, even if in our culture, the positions have been reversed. For there are about 200,000 moth species in the world, as a ballpark figure, but only about 20,000 butterflies: butterflies are just a branch, halfway down, on the moth evolutionary tree, a group of moths which split off and evolved to fly by day, and developed bright colours to recognise each other. This disparity in species numbers is even more pronounced in Britain, where there are a mere 58 regularly breeding butterfly species, but about 900 larger moths (all with English common names) and another 1,600 or so smaller or micro-moths (which for the most part have only scientific names in Latin), for a total of about 2,500. Thus, in the world as a whole, there may be ten times as many moth species as butterfly species; but in Britain, it is approaching fifty times as many.

This means, of course, that in the dark there are far, far more moths out and about than ever there are butterflies during the daytime; it's just that we don't see them. Or at least, we didn't, until the invention of the automobile. The headlight beams of a speeding car on a muggy summer's night in the countryside, turning the moths into snowflakes and crowding them together the faster you went, in the manner of a telephoto lens, meant that the true startling scale of their numbers was suddenly apparent, not least as they plastered the headlights and the wind-screen until driving became impossible, and you had to stop the car to wipe the glass surfaces clean. (I know there are many other insects active at night as well, but let the moths stand proxy for the rest.) Of all the myriad displays of abundance in the natural world in Britain, the moth snowstorm was the most extraordinary, as it only became perceptible in the age of the internal combustion engine. Yet now, after but a short century of existence, it has gone.

In recent years I have often talked to people about it, and I am surprised, not just at how many of those over fifty (and especially over sixty) remember it, but at how animated they become once the memory is triggered. It's as if it were locked away in a corner of their minds, and in recalling it and realising that it has disappeared, they can recognise what an exceptional phenomenon it was, whereas at the time, it just seemed part of the way things were. For example, I talked about it to one of Britain's best-known environmentalists, Peter Melchett, the former director of Greenpeace UK and now the policy director of the Soil Association, the pressure group for organic farming. As soon as I raised the subject he said: 'I remember being at a meeting with Miriam Rothschild [the celebrated natural scientist], and Chris Baines was there, the TV naturalist, the guy who founded the Birmingham Wildlife Trust, and we were talking about the loss of insects in general and the loss of moths in particular, as Miriam was a great moth expert, and I said I remembered in the fifties driving from Norfolk to London with my dad, and him having to stop to wipe the windscreen and the headlights two or three times during every journey, so he could see.' He laughed. 'And Chris Baines said, it was all very well for you, being driven around in a flash car – for me, you couldn't bicycle with your mouth open, because you would swallow so many insects.'

I looked up Chris Baines, and he laughed in turn, and said it was true. He said: 'Yes, I remember it very well, having to scrape the windscreen and the headlights clear of insects, but I did also experience it on my bike. I used to cycle to Cubs or to church choir practice and you would get them in your eye, or if you had your mouth open, you ended up spitting out bits of moth wing, there were just so many in the air on any evening.' He thought about it for a moment and he said: 'If you drove down any kind of hollow way, like a country lane with hedges on both sides, you would be driving through a terrific mass of insects, and now that never happens. I remember it until my twenties. It's difficult to be precise, but I was a student in Kent, at Wye College, and my recollection is that it was still the case then, in the late 1960s, but not after that, really. It certainly never happens now. We spend a lot of time in rural Wales, driving in north Wales, and there have been evenings when I have commented that we've seen a moth. It's almost literally that – one or two moths in a journey. That's a completely different kind of situation from when I was growing up.'

It was in the millennium year, 2000, that I myself began to realise that the moth snowstorm had disappeared, and I began to write about it as part of the issue of insect decline as a whole, which seemed to me to be wide-ranging and extremely serious – the honeybees and the bumblebees were declining, the beetles were disappearing, the mayflies on the rivers were plunging in numbers – but very under-appreciated: no one was interested in it. Yet every time I wrote about the snowstorm, people would respond. They would say how vividly they remembered it, and how now they never saw it, and a frequent memory was of the long drive to the coast for the summer holiday in July or August (in the fifties, Spanish beaches were still in the future) when the car windscreen would unfailingly be insect-plastered; and then it all stopped. The experts remembered it just like the members of the public. Mark Parsons, the principal moth man at Butterfly Conservation, recalled it vividly from twenty or thirty years earlier, but he said to me: 'I may have seen it once or twice in the last decade.'

All this was just anecdotal, of course. There were no scientific figures for moth decline, as Britain's community of naturalists, enthusiastic though they were, had never got round to creating monitoring surveys for moths like they had done for birds, wild flowers, and butterflies. It was merely memories. Then one day the figures suddenly appeared.

They came from an unexpected source: Rothamsted, the celebrated agricultural research station in Hertfordshire (the oldest agricultural research station in the world, in fact, with experiments on the effects of fertilisers on crops going back to 1843). From 1968 Rothamsted had operated, through volunteers, a nationwide network of moth traps, the data from which had been used, within the station itself, to study various aspects of insect population dynamics. But in 2001 it was perceived that one well-known, widespread and common moth, the strikingly beautiful garden tiger, appeared to be collapsing in numbers. As a result, the Rothamsted scientists began to analyse the long-term population trends of 337 larger moth species regularly caught in the traps over the full thirty-five-year period the network had been running, from 1968 to 2002. The results, made public in conjunction with Butterfly Conservation on 20 February 2006, were astounding: they showed Britain's moth fauna to be in freefall. Wholly unsuspected in its scale, the position was even worse than that of the birds, the wild flowers, and the butterflies. Of the 337 species examined, two-thirds were declining: 80 species had declined by 70 per cent or more, and 20 of these had gone down by over 90 per cent. In southern Britain, three-quarters of moth species were tumbling in numbers; their total cumulative decline since 1968 was estimated at 44 per cent, while in urban areas, the losses were estimated at 50 per cent. The snowflakes which had made up the snowstorm were simply no longer there.

It had been the most powerful of all the manifestations of abundance, this blizzard of insects in the headlights of cars, this curious side effect of technology, this revelatory view of the natural world which was only made possible with the invention of the motor vehicle. It was extraordinary; yet even more extraordinary was the fact that it had ceased to exist. Its disappearance spoke unchallengeably of a completely unregarded but catastrophic crash in Britain of the invertebrate life which is at the basis of so much else. South Korea may have destroyed Saemangeum, and China may have destroyed its dolphin, but my own country has wreaked a destruction which is just as egregious: in my lifetime, in a process that began in the year I was born, in this great and merciless thinning, it has obliterated half its living things, even though the national consciousness does not register it yet. That has been my fate as a baby boomer: not just to belong to the most privileged generation which ever walked the earth, but, as we can at last see now, to have my life parallel the destruction of the wondrous abundance of nature that still persisted in my childhood, the abundance which sang like nothing else of the force and energy of life and could be witnessed in so many ways, but most strikingly of all in the astonishing summer night display in the headlight beams, which is no more.

But if we know full well why half our wildlife has gone – step forward, Farmer Giles, with your miserable panoply of poisons – the reason for the disappearance of one particular part of it, London's sparrows, remains a mystery entirely.

How utterly bizarre that it should happen to him, the Cockney sparrer! The urban survivor par excellence! The bird that has lived alongside humans since human settlements began twelve thousand years ago . . . the bird which is wholly at home in the city . . . what is it that, in one of the world's greatest cities where previously it flourished, has destroyed its population? To this day, more than twenty years after the event, nobody knows.

The phenomenon is all the more perplexing in that in major cities ostensibly very similar in infrastructure and atmosphere to London, such as Paris or New York or Washington, sparrows are flourishing still, darting in their flocks around the feet of the tourists in hope of the dropped crumb or the piece of ice-cream cone. Yet in Britain's capital, over the decade of the 1990s, the population collapsed, and the birds vanished almost completely. Within the London sparrow ecosystem, something mysterious, something catastrophic, took place. But even now, no one has worked out what.

The bird we are talking about is the house sparrow, *Passer domesticus*, which has hitherto been one of the world's most successful creatures. It occurs naturally all across Europe, much of Asia and North Africa, and has been introduced to Southern Africa, the Americas and Australasia: Antarctica is the only con-tinent without it. It has been found breeding at 14,000 feet up in the Himalayas and nearly 2,000 feet down in Frickley Colliery near Doncaster (really: in 1979). It is one of the world's commonest birds, and almost certainly the most widespread; but more than that, it is beyond doubt the most familiar. Down the ages, the house sparrow has generated a special affection in us, based on its close association with people and towns, and a perception of its character as humble but hardy; as an urchin, but an urchin that lives on its wits. When Hamlet told Horatio there was a special providence in the fall of a sparrow, he was making it the exemplar of the lowly, but the bird was that already, more than sixteen hundred years earlier in Rome: Catullus' famous and charming poem on the death of Lesbia's sparrow is mock-elegiac, calling all Venuses and Cupids to grieve for his lover's beloved pet. Lowly, yes; but also street smart, like Paris's most celebrated singer, tiny and irrepressible, who called herself after the French slang word for sparrow, *piaf.*

The house sparrow has needed its survival skills. When I asked the world expert on the bird and on sparrows in general, Denis Summers-Smith, what he liked most about them, he took me by surprise; he said: 'I greatly admire their ability to live with an enemy.' 'Who's the enemy?' I said. 'Man,' he said. I said I thought that sparrows and humans had always got along fine, but he disabused me of that; farmers in particular used to hate them for the grain they consumed, he said, yet the birds continued to live on farmhouses. They were often killed, but they managed to get by, from generation to generation, by remaining intensely wary of this primate with whom they had thrown in their lot. Speaking of when he first began observing them closely, from his Hampshire garden in the late 1940s, Denis said: 'If I was gardening, they wouldn't look at me, but if I started to look at them, then they would look at me. They were very conscious of me. If I was going about my normal business, they weren't bothered, but as soon as I started watching them, they would watch me back.'

A Scottish engineering consultant and former senior scientific adviser to ICI, who at the time of writing is ninety-three and still going strong, Denis has been studying the twenty-seven members of the genus *Passer*, and *Passer domesticus* in particular, for nearly seventy years, a lifelong interest which has made him perhaps the most eminent amateur ornithologist in Britain of the second half of the twentieth century, with five books on sparrows to his name, including the standard monograph, *The House Sparrow*, published in the famous Collins New Naturalist series in 1963. In them, he elucidated many aspects of sparrow private life which yet may have a bearing on its mysterious London collapse; two in particular are that sparrows are very sedentary, and sparrows are very social. They are in fact the most sedentary of all songbirds, usually living out their lives within a one-kilometre radius, and foraging if they can within fifty metres of the nest; and their sociability is just as pronounced. Sparrows live in colonies: they deeply need and depend upon each other. This is vividly illustrated by the behaviour Denis has christened 'social singing'. After feeding, with their crops full of seeds which need time to be digested, sparrows gather in cover such as a thick bush, in groups of typically a dozen, and sit back, as it were, and begin cheeping to each other. The call generally sounds like a monosyllabic *cheep*, although if you slow it down, it is clearly a disyllabic *chirrup*! They each take it in turns to give a single sound, with a separated abruptness which is very distinctive:

> Hey!
> What?
> You!
> What?
> You!
> Eh?
> Who?
> Him.
> *Him?*
> Nah.
> *Her?*
> Nah.
> Me?
> Nope.

Him?
Yup.
Really?
Yup.
Me?
Yeah.
Oh.
Yeah.
Why?
What?
Me.
Cos.
What?
You.
Eh?

This was one of the most familiar sounds of my childhood in the suburbs, when sparrows were everywhere; it is almost wholly lost from London now, even though comparable small songbirds, from robins and wrens to blue tits and blackbirds, continue to give full voice in the capital's parks, and the other archetypal bird of the city, the feral pigeon, prospers as ever in London's streets (and makes up most of the diet of the peregrine falcon, several pairs of which now breed in the heart of the capital). What was different about the house sparrow, that it was singled out for disappearance?

Certainly, there had been an extended decline through the length of the twentieth century: the figures are there. In November 1925 a young man of twenty-one went into one of central London's greenest parks, Kensington Gardens, and with the help of his brother counted the house sparrows: there were 2,603 of them. The man was Max Nicholson, a passionate ornithologist and the founding father of Britain's environmental institutions, who as a senior civil servant in 1949 brought into being the world's first statutory conservation body, the Nature Conservancy, and subsequently ran it for fifteen years; he ended up as the Grand Old Man of the natural world in Britain, having been the founding secretary of the British Trust for Ornithology, president of the Royal Society for the Protection of Birds, and having helped to launch, in 1961, the first of the world's great Green pressure groups, the World Wildlife Fund (now the Worldwide Fund for Nature). But for all his prominence in officialdom, Nicholson remained a practical ornithologist at heart, and in December 1948 he repeated his Kensington Gardens sparrow survey: there were then 885 birds. In November 1966 there were 642, and in November 1975 there were 544; but when he took part in the count in February 1995, at the age of ninety-one, there were a mere 46, and on 5 November 2000 I went back with him to Kensington Gardens – he was ninety-six by now – and we watched as members of the Royal Parks Wildlife Group carried out a seventy-fifth anniversary census of his original count: they found 8 birds.

What on earth had happened? The earlier decline apparent in the Nicholson figures, between 1925 and 1948, has been attributed to the disappearance from London's streets of the horse, and the loss of the grain spilled from nosebags and even the undigested grain in horse manure, which was an important source of food for small birds; but then for forty years or more the sparrow population was on what we might call a gently declining slope. However, from about 1990, it fell off a cliff: this is the enigma. In Buckingham Palace gardens, which in the sixties supported up to twenty sparrow pairs, there were none after 1994; and in St James's Park, where once sparrows could be found by the hundred, where squabbling flocks of them would cluster on the shoulders and arms and palms of bird-food-proffering tourists – I can remember that myself – a single pair nested in 1998, and in 1999, for the first time, no birds bred.

Alert observers began to notice. Among the first was Helen Baker, then the secretary of the ornithology research committee of the London Natural History Society, whose morning walk to work at the Ministry of Agriculture in Whitehall took her through St James's Park. In particular she noticed that the sparrows had gone from the shrubbery at the end of the bridge over the lake, where in the past she had counted the birds by the hundred and had had them feeding from her hand; and in 1996 she organised the LNHS house sparrow survey, to try and get a handle on what was happening. News of the decline began to seep out in London's evening paper, the *Evening Standard*; I myself became aware of it in 1999, realising that the sparrows had gone from my commuter terminus, Waterloo Station, where once they had been plentiful. I began to look out for them, and couldn't spot them; but it was not until a trip to Paris with my wife and children in March 2000 that the true scale of the situation dawned on me, for in the French capital *les piafs* were everywhere, in stark contrast to London, where now they seemed to be nowhere. I wrote a piece about it which was featured prominently in my newspaper, the *Independent*; I continued writing about it; and eventually in May 2000 we launched a campaign to Save The Sparrow, the centrepiece of which was a £5,000 prize for the first scientific paper published in a peer-reviewed journal which would explain the vanishing of the house sparrow from London and other urban centres, in the opinion of our referees, who were the Royal Society for the Protection of Birds, the British Trust for Ornithology, and Dr Denis Summers-Smith.

The *Independent*'s campaign, and especially the offer of the £5,000 prize, put the disappearance of London's sparrows firmly on the news agenda nationally and internationally – it was reported around the world – and it elicited a substantial reader response, with nearly two hundred and fifty letters in the first weeks (about twenty of them being emails; this was just on the cusp of the email revolution, and most of the missives were still handwritten or typed). There were two significant aspects to this. One was the surprising passion with which people lamented something so seemingly inconsequential as the disappearance of a small brown bird: it was as if an emotional floodgate had been opened, a commonly expressed feeling being gratitude that someone besides the writer had at last taken note of this development and also considered it important ('I thought it was only me . . .').

The other prominent aspect to the response, of course, consisted of readers' theories for the disappearance, and two weeks after launching the campaign, we listed ten of them. They were, in order of frequency of expression: predation by magpies; predation by sparrowhawks; predation by cats; the effect of pesticides; the tidying up of houses and gardens, which removes nesting places; loft insulation, ditto; climate change; the effects of radiation from the Chernobyl disaster; the introduction of lead-free petrol to Britain in the 1990s; and finally, peanuts (the suggestion being that the vogue for putting peanuts in bird feeders was perhaps upsetting the sparrows' digestion – fatally). The response reflected the detestation of the British bird-feeding classes for the magpie, the bold black-and-white crow which from the 1970s onwards had moved from its previous rural habitat into suburban and urban gardens – the sparrowhawk effected a similar shift in the 1990s – and was often observed preying, with upsetting relish, on songbirds, their nests and eggs and chicks. Nearly all the letters were deeply felt, although the odd one was a tad presumptuous ('It's cats. Send money to address below . . .').

But if deeply felt, they did not necessarily reflect expert judgement, so I sought out the experts. I went to see the venerable Max Nicholson in his house in a backwater of Chelsea, and in his curious high-pitched lisp, with an articulacy quite undimmed by the imminent approach of his ninety-sixth birthday, he advanced what at first seemed to me to be a quite startling idea: that, in his words, sparrows as a species had a strong suicidal tendency. What he meant was that if sparrow numbers, in the colonies in which they nested, fell below a certain level – for reasons such as a lack of food – the colony might suddenly cease breeding and dissolve. The problem, he thought, was ultimately a psychological one: the birds, which were so strongly social, felt that life in such low numbers was no longer worth living. The basis of this idea is actually supported by a well-known piece of biological theory, the Allee effect, which states that declines in socially breeding species can become self-reinforcing, but it was the vividness with which Max Nicholson expressed it which initially took me aback. 'I think they suddenly get to a critical point where they say, let's give up,' he said. 'I don't think it's about safety in numbers. I think it's psyche.' He said that this should be correlated absolutely with material factors like food shortage, which was his own suggested trigger for the initial numbers drop which might precipitate a psychological crisis; and he stressed he was speculating, and fully accepted that what he was suggesting would be difficult to verify experimentally. 'I accept it's an element that can't be measured,' he said. 'It's a psychological thing – there's no scientific way of measuring it.'

He smiled.

'But a lot of things that can't be measured, are real.'

I thought then and I think now that Max Nicholson may have been right and that sparrow colonies might well drop in numbers to the point where they suddenly dissolved; but the mystery was, what was causing the drop? And Denis Summers-Smith, when I went to see him at his home in Guisborough in the north-east of England, had a specific view about that. With his intimate knowledge of sparrow biology, he was aware that although sparrows are granivorous birds – they feed on seeds – the sparrow chicks, for the first few days of their lives, need insect food, such as aphids (the greenfly abhorred by gardeners), small grubs, flies, and spiders. He conjectured that insect numbers might have fallen, and to the point where the chicks might starve and the birds' reproductive rate might fall itself, triggering a population decline, since to make up for natural winter mortality and maintain their population levels, sparrows need to rear between two and three broods of chicks every summer.

And Denis had a candidate for what might be killing off the insects in towns and cities like London: motor vehicle pollution, and specifically, the introduction of lead-free petrol into Britain, in 1988; for not only did that represent the major change in the composition of vehicle exhaust fumes in previous years, but there was also a strong temporal correlation between the introduction of unleaded and the sparrow decline itself (as a couple of canny readers had noticed, in our initial trawl of ideas). At first, unleaded sold in only tiny amounts, but sales picked up rapidly during the nineties, leading to the complete phasing out of leaded petrol at the end of 1999, and the uptake clearly paralleled the London sparrows' demise. It was the substitute chemicals added to the petrol to replace the lead and reboost the octane rating, Denis believed, which might be causing the problem, and he focused on two additives in particular: benzene and MTBE (methyl tertiary butyl ether), both of which had health and safety question marks against them. He accepted there was no scientific evidence as yet linking MTBE or benzene directly with house sparrows, but he thought that the circumstantial evidence of a connection was strong. Hence he took the view: 'This is my hypothesis – what's yours?'

It was intriguing, and a potentially devastating example of the law of unintended consequences. Unfortunately, it was a hypothesis that was very hard to test, as although a highly specialised agricultural research station like Rothamsted might be uniquely equipped to measure insect biomass on farmland, nobody at all, as far as I could find out, was measuring insect biomass in towns and cities; it was seen as a near impossible job, and anyway, what would be the reasons to fund it? So you simply couldn't tell if the aphid population of St James's Park, say, was plummeting. I also felt there was a gaping hole in the theory: New York and Washington had unleaded just as London did, and Paris had *sans plomb* – so why weren't their sparrows disappearing in the same way?

Yet Denis's instinct that the proximate cause of the decline might be starvation of the chicks through lack of insects was eventually borne out by a young postgraduate research student at De Montfort University in Leicester, Kate Vincent. For her doctoral thesis, Kate put up more than six hundred sparrow nest boxes in the Leicester suburbs and the adjoining countryside, and monitored them for three years, closely observing the birds' breeding success. (I visited her and watched her gamely clambering up and down her ladders.) Her finding, in 2005, was remarkable: that in the summer, completely unseen by the outside world, considerable numbers of sparrow chicks were starving to death in the nest, and the closer towards the centre of town the nest was, the higher the mortality. Furthermore, those whose diet had consisted largely of vegetable matter – seeds and scraps of bread – were much more likely to die than those whose diet had contained plenty of invertebrates. (Kate worked out the chicks' diet by analysing their droppings: in an ornithological labour of Hercules, every time she weighed and measured a chick in the nest, she collected the poo it would tend to deposit in her hand, and then, under the microscope, she could identify in it the tiny remains of insects – an aphid leg here, a beetle mandible there – and estimate their abundance.) The chicks that were dying were largely in the sparrows' second brood of the year: Kate found an 80 per cent success rate in the first brood, but only a 65 per cent success rate in the second, and with the birds needing between two and three broods annually to maintain their population levels, this could be enough to precipitate a decline.

Eventually, Kate wrote up her findings in a scientific paper with fellow researchers from the RSPB and English Nature (then the government's wildlife agency), and in November 2008 this was entered for the *Independent*'s £5,000 prize. However, the referees were split. The problem was that Kate's research revealed the starvation but not why the insects were hard for the birds to find: it was half a solution. One referee said, award the prize. One said, do not award the prize. And the third said, award half the prize. In the circumstances, it did not seem possible to award it. And there, to date, the matter rests.

In early 2014 I went to Guisborough again to see Denis Summers-Smith and talk the whole issue over once more, fourteen years after we had first highlighted it. I spent two enjoyable days admiring his wonderful sparrow archive of more than five thousand items, and his collection of sparrow artefacts ranging from Chinese sparrow fans to Japanese sparrow netsuke, and we talked late into the night of such subjects as, what species was Lesbia's sparrow? (Denis thinks it was the Italian sparrow, *Passer italiae*, which replaces *Passer domesticus* in the Italian peninsula, although the Spanish sparrow, *Passer hispaniolensis*, also occurs in southern Italy. My friend, the academic ornithologist, Tim Birkhead, basing his view on the sound it made – 'pipiabat,' says Catullus, 'it used to *pipe*' – thinks it was probably a bullfinch.) And Denis told me of how his involvement with sparrows had begun, which really dated back to 6 August 1944, when he was a twenty-three-year-old captain leading his company of the 9th Cameronians in Normandy, in the race to close the Falaise gap, and a German shell landed by him and almost took off his legs. But not quite: eight operations later he still had them, and lying in hospital in Worcestershire, he became fascinated by the sparrows which came in through the windows of the ward. When he had recovered (although with legs full of shrapnel that set off airport alarms), he began his lifelong study.

He had changed his mind about unleaded petrol and MTBE, although he still believed motor vehicle pollution was to blame for the decline of sparrows in London and other urban centres; now, however, he thought that a major cause of the decline was 'particulate' contamination from diesel engine exhausts (essentially nanoparticles of soot that are not filtered out in the nasal passage). This may have led directly to mortality of juvenile birds, he thought.

For my part, I wanted to discuss with him the question that continued to preoccupy me: how could the house sparrow have been 'singled out', as it were, for disappearance? How could it vanish from St James's Park, say, when similar songbirds such as robins and blue tits, blackbirds and wrens, still seemed to lead satisfactory lives there?

The key fact, Denis said, was that sparrows did not disperse.

I asked him what he meant.

He said: 'They live in a small area, which they get to know very well. They spend their lives within a kilometre. They are completely sedentary, the most sedentary of all passerines. But other small birds, like blue tits or chaffinches, are unable to do this; when they leave the nest, they have to disperse. They have to move considerable distances away, to find food or new partners.'

And how did that relate to the situation in St James's Park?

Denis said: 'If the sparrow population in St James's Park dies out, it will not renew itself, because no new birds will come in. But if the blue tit population dies out, other young birds, which are dispersing, will arrive.'

The implication started to dawn on me.

I said: 'So is it possible, then, that what went wrong in the ecosystem . . . what made the sparrows die out . . . is actually affecting all species? But the other species, because they are dispersers, are able to renew their populations . . .'

Denis said: 'Yes.'

'But we can only observe the effect in sparrows, because the sparrows are the ones that can't replace themselves . . .'

Denis said: 'This is my hypothesis.'

'So what we may actually be looking at is a disguised devastation of all these common species?'

'Yes.'

I was dumbfounded. 'This is completely new, Denis. Nobody's ever said this.'

'Well I've been saying it to a lot of people.'

Was it possible? That *all* the birds of St James's Park died out, or failed to breed successfully, every year? But all, except the house sparrows, could renew their populations from outside?

That we had actually witnessed something far more wide-ranging than the downfall merely of *Passer domesticus*?

I could not say.

Whatever had done it so effectively to the sparrows, and possibly was doing it without our knowledge to all the songbirds of central London, and possibly even to more organisms than that – including us – remained unknown.

We still had no idea what it was.

I come from the north of England, but I have lived in London for forty years and grown to know it well and love it, and when I first realised the sparrows had gone from its heart, I felt the loss as keenly as other people did. And six months after visiting Denis, while writing this book, I was suddenly seized with a desire to go out into central London and look for them, wondering if, two decades after their disappearance, any trace of them might remain.

I approached Helen Baker, who had spotted the birds' disappearance almost before anyone else. She had risen in the London Natural History Society and was now its president, but was as fascinated as ever by the sparrows' fate; she was a receptacle for all the reports which surfaced from time to time, of the odd colony clinging on here and there, in quiet corners. Helen told me she thought there might be three small colonies still in central London, two of them on the South Bank, and on a hot July day we set out in search of them, meeting in the Guildhall Yard, the very hub of the old City. Helen was attending a lunchtime concert in the Guildhall church and while I waited for it to finish I watched the office workers with their sandwiches being half-heartedly hassled for crumbs by the pigeons. When I first came to London, sparrows would have been the hasslers-in-chief.

We began our South Bank search at Borough Market in the shadow of Southwark Cathedral, whose pinnacled tower Shakespeare would have eyed (the churches on the north bank all being consumed, of course, in the Great Fire of 1666). Borough Market epitomises what we might call the Mediterraneanisation of London which has taken place over recent decades – the introduction into the capital of exhilarating new foods and the enthusiasm of crowds for them and habits of eating in the open air (on a sunny day it could almost be Barcelona) – and if ever there was a place where sparrows would thrive, this was it. The local birds knew it too. But they were pigeons and lesser black-backed gulls and, I was delighted to see, a crowd of starlings; of *Passer domesticus*, there was no sign. There was no sign of him either as we skirted the replica of Francis Drake's *Golden Hind* in St Mary Overie dock and moved up Clink Street past the medieval remains of the Bishop of Winchester's Palace and out on to the riverbank, and the Anchor Pub. In the garden beyond the old pub, said Helen, sparrows had occasionally been seen over the previous year, and we watched and we listened for several minutes, because with sparrows you may well hear them cheeping before you see them. The only sound was the laughter of drinkers. There were no sparrows there that day.

Helen's second potential South Bank sparrow site was another garden, further upstream at Gabriel's Wharf, and as we walked there I was struck by the number of pigeons, especially outside Tate Modern, the power station turned temple of contemporary art on whose soaring art deco brick tower peregrine falcons – notable pigeon consumers – roosted. 'It pleases a lot of people that peregrines eat pigeons,' said Helen, who explained that in the school holidays she was one of the people manning the RSPB telescope trained on the tower so that the public could observe the peregrine pair who had been named Misty and Bert. I wondered how many of the pigeons I was watching would end up as peregrine dinners; they were in the sort of numbers that sparrows would once have exhibited, hundreds and hundreds of them. But there was no sign of the sparrows; not there, not anywhere along the embankment, and not at Gabriel's Wharf either, where we thoroughly explored the garden in which Helen had in the past counted up to forty; now there were merely sixteen pigeons on the lawn. 'Oh, this is disappointing,' said Helen. 'It used to be such a very good colony. It may be that the food supply has gone. They used to nest in the houses and flats nearby. One would hear them and one would see them, going back and forth from these houses.' Not any more.

It seemed to me that London was completely sparrow-free; for the South Bank was such a tourist trap, it had so many eating places with people sitting outside dropping crumbs, that in any other European city it would be a sparrow food resource par excellence. But there were none whatsoever. It was uncanny. It was chilling, almost. The disappearance of the birds seemed complete.

Helen had one remaining site to try, which was on the north side of the river, so we walked over Waterloo Bridge and into the West End; we wound our way into a celebrated and historic area, and Helen said to be on the lookout, for birds had been seen in the street we were in, on the window boxes. 'Look up, keep your eyes open,' she said. I could see nothing. We turned into another street, a famous one, and she repeated her exhortation; I could still see nothing. Then, as we were passing a well-known Italian restaurant, I heard it:

Hey!

What?

You!

What?

You!

Eh?

Who?

Him.

Him?

Nah.

Her?

Nah.

Me?

Nope.

Him?

Yup.

Really?

Yup . . .

A flood of elation swept through me. I shouted: 'I can hear them! I can hear them!' Helen called out: 'I can see them too!'

'Where?'

'Here they are on the wall . . .'

'Oh God, yes! Suddenly! Two of them!' – all this from my tape recorder – 'Wow you're right! A third one! On the flats, on the old Victorian flats!'

They might have been the rarest birds in the land, red-backed shrikes or black-winged stilts, they might have been Siberian rubythroats, such was my delight. I said to Helen: 'I never imagined I would ever feel this way about sparrows.'

The chirping was continuous by now. We were opposite a tiny park, just a garden really, full of bushes: the chirping was coming from inside, and when we went in, we found the birds, hovering around feeders which had been placed deep into cover. It was in a very quiet part of a famous street, almost a backwater in the heart of tourist London; the birds foraged in the garden, and nested in the old flats across the road.

Just a handful of them.

Very shy, hiding in the foliage.

But there they were.

Finding a tiny colony of house sparrows in central London does not make up for losing the whole population; but it does something. It's a smidgeon of light, I suppose. Sometimes I think there is no light; but sometimes I think there is. For we have the losses, the losses which are now so extensive and ruinous they are coming to define the natural world, the losses which are wrecking the earth and its biosphere to an extent hitherto unimaginable, the losses which are making us seem, as a species, like a curse, like a blight upon the fragile, exquisite, isolated planet which is our only home; but we also have the bond.

Our bond with nature may be hidden for much if not most of the time, it may be a signal engulfed by the noise, it may lie buried under five hundred generations' worth of urban living, but it is stronger than those experiences, for it was forged by fifty thousand generations of living in the natural world before the farmers broke the sod and hacked down the forest and imposed a new order on humankind; and underneath everything, it endures. It is unbreakable. Nor does it belong just to him, or to her; it is the inheritance of every single one of us, it is part of what it means to be human, and it can be found within us – not always easily – and it can be understood, and it can be made the basis of our defence of the natural world in the terrible century to come. So let us leave them behind, the unbearable losses, and go where the bond can be found: let us journey into joy.

5

Joy in the Calendar

It would be foolish to underestimate, however, the obstacles in the way of finding and feeling our inherent bond with nature, which will grow substantially as the century progresses; that needs to be admitted. At some unknown moment between 1 July 2006 and 1 July 2007, according to the demographers of the United Nations, a momentous milestone in the history of humankind was passed with no one being aware of it: the percentage of the world's population living in towns and cities exceeded 50 per cent. Henceforth, most people on the planet would live urban rather than rural lives, and for the first time would no longer be in close contact with the natural or even the semi-natural world (which farming represents); a majority, and a rapidly expanding one, would no longer have direct access to the rhythms of the growth cycle, to the effects of seasonality, to quiet, to the visibility of the stars, to non-industrialised rivers and natural forests, and to wildlife – to birds and wild mammals, to insects and wild flowers – even where, as in more and more places, wildlife was impoverished. Nature in any form would no longer be part of most people's everyday experience.

It is worth looking for a moment at just how quickly the

urbanisation of the globe is now proceeding. In 2014 the proportion of people dwelling in towns and cities reached 54 per cent, according to that year's revision of the UN's *World Urbanization Prospects* (this document redates to 2006–7 the passing of the 50 per cent mark, previously thought to have taken place in 2009), and this figure is expected to increase to 66 per cent by 2050: that is, 6 billion out of an anticipated 9 billion souls, or two-thirds of the world.

Nearly all of this increase – 90 per cent – is expected to take place in Africa and Asia, much of it in their 'megacities', the mushrooming metropolises of 10, 20, 30, going on for 40 million people which will be one of the most notable facets of human geography in the twenty-first century: by 2030 the world is expected to have forty-one of them. Running these gargantuan settlements, and the hundreds of 'smaller' cities which will unstoppably expand to a million-plus, 3, 5, 7 million people and more, will present the greatest social and infrastructural challenges, in providing adequate water, food, healthcare, education, transport, energy, employment, and housing. You can take a positive view. A case can be made for cities, and often is, even for the megacities: they can generate jobs and income, and can deliver health and education and the empowerment of women, say, more efficiently than can be done over vast rural areas (as long as the cities concerned are well governed). But these are issues which people involved in human welfare, in poverty and its alleviation, are concerned with, and rightly so; I am concerned with the natural world and the human response to it, and I cannot see how that will be benefited in any way by what we might call the great urban shift.

Instead, nature may come to represent for billions – for two-thirds of the world, by mid century – merely what the city is not: a folk memory of clean air rather than smog, of clean rivers rather than polluted ones, of grass and trees rather than concrete and cars, of wild creatures freely existing, now seen merely in visual representations. That will be entirely understandable; and if swelling urban environments continue to intensify the stress and the pollution they inflict upon their residents, we could not but wish city dwellers the world over the chance of escape to the trees and the grass, the pure water and the pure air, which are worth so much, especially when we mix our pleasure with them and find it so enhanced, in the picnic by the riverside or the ramble through the forest . . . but something else too, in the great shift, will be lost.

It is the intimate feel for the natural calendar, for the earth's great annual cycle of birth and death and rebirth, a feel which was one of the key attributes of our prehistoric ancestors and which has persisted among people living in the countryside long after city dwellers lost the conscious sense of it. Not lost quite entirely, of course: even a geek working in the most concrete-and-glass-bound thirtieth-floor neon-lit air-conditioned cappuccino-dispensing digitised electronic bolt-hole will sense it is hotter in summer and cooler in winter – but I mean something subtler. I mean the feel for the switches and the transformations, for the tiny signs, easily stifled by traffic noise and electronic music or submerged by pollution, that great changes are under way with the earth; the feel for the hints of the journey starting, rather than the trumpeted proclamation of the arrival. These signals, above all of the world's reawakening after winter, have produced intense pleasure and excitement and indeed reverence in us since we began to be human, they have produced the most powerful emotions, and not infrequently in my own case, they have produced joy.

Journeying into joy, it is where I would start from. And the loss of this, the loss of familiarity with the cadences and pulses of nature which will extend to so many more of us in the two-thirds urban world of the years to come, seems to me to be sad beyond words, not least because it will go unmarked and unmourned, since for someone struggling for food and basic healthcare and education for their children in a megacity shanty-town without sanitation or energy supplies, that will be the most minuscule of their concerns. The rhythms of nature? It will be no sort of concern at all. Of course. And yet it is a great loss nonetheless, as I increasingly feel looking over the joy I have indeed encountered there, the joy I have found in the calendar and in the signals of the awakening world, beginning with the winter solstice.

The winter what? many people will say, especially young people. Believe me, they will. And this, the most significant moment of the year! The moment when the days stop shortening and start getting longer again, celebrated for millennia. It is the reason (or one of the reasons) for Stonehenge; it is the reason for Newgrange, Ireland's premier prehistoric monument; it is the reason for Christmas in December. (The Wiltshire megaliths are lined up to the winter solstice sunset; the grand tomb in County Meath is aligned to the winter solstice sunrise; 25 December was chosen by the early Christian church as the conventional date of Christ's birthday since it was the date of the winter solstice in Roman times.) Now it has shifted, partly because of the replacement of Julius Caesar's Roman calendar by the Gregorian calendar from 1582 onwards, and it occurs on 21 or 22 December. It is not actually a day, but a precisely calculable moment in the earth's orbit when the tilt of its axis is farthest away from the sun: thus in 2010, for example, it was 11.38 p.m. on Tuesday, 21 December, but as it had occurred after sunset, the celebration, as such, was held the day afterwards.

Not by many of us, though. A diverse band of druids, pagans, hippies, and sundry sun-worshippers gathered at Stonehenge that day to mark the moment with rituals and dances, as is their wont, providing useful colour for the news media; but otherwise, the modern mass of humanity got on with their lives while paying the most significant day of the year scant attention. It is the archetype of the momentous marker that we have forgotten, the winter solstice, in our harried urban existence where we don't see the stars for the street lights and never notice the sunset — we do that on holiday, don't we? *Darling? Come and see the sunset!* — and certainly not three days before Christmas when everyone is feverishly preparing for the winter break.

Yet as I have got older, I have come to love it. For whether or not in our flurry of living we lose touch with the rhythms and processes of the earth, behind everything they continue in all their power, and the solstice represents the start of the most powerful of them all: rebirth. The moment when the days begin to lengthen again is the moment when new life begins its approach, even at the darkest point, which is why it has been so widely celebrated in so many cultures right round the world — the miracle of rebirth never ceased to amaze. Death was being refuted. It was wondrous that new life should arrive quite as unfailingly as old life should die, especially since a human individual's life itself was linear — it only went in one direction. But the earth was different. Its way was not linear; it moved in a cycle, and although you might fear that one year the cycle would break down, it never did.

Ageing has made me more appreciative of the miracle (partly, I suppose, from a rueful recognition that it isn't going to happen to me) and what that has produced has been a heightened awareness of its advent, even though each year, for most people, it may lie buried deep under the pre-Christmas frenzy of packed stores and heaving parties and crammed buses and suffocating trains and chaotic airport terminals — but if you do take the trouble to look closely, you will see that behind all the craziness, it is happening. Going back to 2010, for example, on Christmas Eve, Friday, 24 December, had you broken off briefly from your last-minute panic present-buying, you would have seen that sunset was at 15.55, but the following afternoon, when the world was recovering from Christmas lunch, it was at 15.56, while on the Monday, 27 December, when some people were drifting back to work and some people weren't and others were wondering whether they should or not, it was at 15.57; and by New Year's Eve, Friday, 31 December, with everyone getting ready for the final seasonal splurge, it had broken the four o'clock barrier: it was at 16.01. And so it goes on each year, in these wholly unremarked-upon, these virtually imperceptible yet remorseless gradations, until it's some time at the end of the first week of March, say, and you're home from work early for some reason, maybe you went to the dentist or something, and you look out of the kitchen window at about ten past six and you notice it's still light, and the light has a special calm but intense quality to it, and this is something different, something new: it's evening. Evenings are back. And you open the kitchen door into the garden and a blackbird is singing from the roof opposite, and a song thrush from the tree next door, both of them liquid and loud and confident in this new-found radiance, in a moment in time which those who have experienced it will realise, Philip Larkin captured with exquisite perfection:

> On longer evenings,
> Light, chill and yellow,
> Bathes the serene
> Foreheads of houses.
> A thrush sings,
> Laurel-surrounded
> In the deep bare garden,
> Its fresh-peeled voice
> Astonishing the brickwork . . .

and you suddenly realise that the whole world is on the tremulous verge of something immense: spring is coming.

The winter solstice is the beginning of that. Friends have said to me their least favourite months are January and February, but I've never thought so; I've always least liked November and early December, when the movement of the earth is only down towards the dark. The first two months of the year may be harsher in terms of weather but ticking in the background is the wondrous phenomenon, the unstoppable movement back towards the light, and for marking its onset, I have got to the stage now where I look forward to the winter solstice more than Christmas, which so swamps and dominates our culture. Not that I have anything against Christmas itself: having been brought up in the Christian fold, I have reverence for its story and enjoy its customs and music and celebrations, in the way that you can if you've been lucky enough to have had them refreshed for you through children, even though they are so naffly commercialised, and even though I recognise that for some people the whole business can be a hateful period of truly glacial isolation.

But the solstice . . . I can only say that, as I move towards the last part of my life, its arrival fills me with joy (even if I don't trek out to Stonehenge), in the way I tried at the outset to define joy in the natural world: a sudden intense love stemming from an apprehension that there is something extraordinary and exceptional about nature as a whole. I can think of nothing more extraordinary and exceptional than the annual rebirth of the world; and in fact, there are a number of specific markers of the rebirth, of the earth's reawakening after winter, dates in the natural calendar if you like, which for me are occasions of joy almost as much as the solstice is, and which I celebrate in my heart.

The first of them is the appearance of snowdrops. Small white lilies which sprout up and flower in midwinter, even in the bitter cold – *perce-neiges*, snow-piercers, they're aptly called in French – would be notable anyway, but I've long been equally fascinated by snowdrops for their cultural resonance. They are closely associated with a major feast of the Christian church which follows Christmas, although while the world and his wife cannot remain ignorant of 25 December, indeed, cannot get out of the way of the Yuletide juggernaut, I doubt if one person in a thousand could tell you today what Candlemas is.

Celebrated on 2 February, it marks the purification, under Jewish religious law, of the mother of Christ, forty days after his birth. (It also commemorates the presentation of the infant Jesus in the temple.) But Candlemas long meant something else as well, in practical terms, especially in the Middle Ages: it was the day when everyone in the parish brought their candles to church to be blessed by the priest. This was so that they could become − that splendid word − apotropaic, that is, they could ward off evil spirits; and after a procession, and the blessing, the candles were all lit and set before the statue of the Virgin Mary. Imagine: on a typically murky February day, in a medieval church that was gloomy anyway, this must have provided a spectacle of brightness that left the deeply pious onlookers spellbound; it must have been the brightest moment, quite literally, of the whole year. (You can get a feel for it if you visit Chartres, and come across the luminous flickering throng of candles in front of the Virgin's statue in one of the cathedral's darker corners.)

But another source of brightness was also closely associated with Candlemas, and that was the snowdrops, for they were the flowers of the feast. It is easy to see how they were perfect for it, flawless symbols of purity that they are. Once called Candlemas bells, it is not hard to imagine what pleasure must have been taken in gathering them, or in merely having them growing by the church, on the day itself; and even now, although you can find great swathes of snowdrops in woodlands or along river valley floors, especially in the West Country − stirring sights, whole sheets of blooms turning the ground white in all directions, nature with all its flags flying − many of our best displays are still associated with the old faith, clustering around church-yards and ancient religious foundations, ruined abbeys and priories, where hundreds of years ago they were planted with Candlemas in mind.

All of this has greatly drawn me to them, yet even more than their delicate beauty, more than the traditions which cluster around them, I am most taken with the timing of their appearance, with their place at the start of the calendar – above all, with the first sight of them in any given winter. I can remember, for example, a walk with my children, a few years ago, through a wood on an icy late January day, and the path through the bare trees took a turn and suddenly there they were, the first of them, a small clump poking through the leaf litter, a small splash of brilliant white on the woodland floor's dull brown canvas, and I smiled at once, as if suddenly meeting an old friend: *Hi, how are you?* I was filled with emotion; I was filled with joy, I would say now. I wasn't quite sure then why the feeling was so strong, but that evening I sat down and worked it out: here was the earth, still firmly under the lock and key of winter; here was I, huddled inside my coat, adjusted to the cold hard season as if it would last for ever; and here were they, the first visible sign of something else. They were the unexpected but undeniable notice that the warm days would come again, and I realised what it was that made me smile: here against the dead tones of the winter woodland floor was Hope, suddenly and unmistakably manifest in white.

Snowdrops are singular. They alone are the flaunters of this optimism, which can seem gloriously defiant, in the heart of the time when the earth is anaesthetised and numb. But as the world starts to stir again, to wake, to warm and to open, there are an increasing number of signals of spring, for some of which my feeling is so intense that I would readily describe it as joy. One is the appearance of the first butterfly, especially if – as is often the case in Britain – that butterfly is a brimstone (which, being bright yellow, the colour of butter in fact, is perhaps the origin of the *butterfly* term – or perhaps not . . . nobody really knows). This event has on occasion had a peculiar effect on me: it has produced an elation so powerful that I have found myself longing for an unconventional way to account for it, to do it justice, in the conscious knowledge that what is available – science – is inadequate for the task.

It has been well said, that science gives us knowledge but takes away meaning. Certainly, since it began to explain the world in rational terms in the seventeenth century, it has subverted or done away with many parts of our imagination, and there are numerous non-rational ways of looking at the world, once widespread, once resonant traditional beliefs, which we have now ceased to engage with, such as alchemy, or magic, or the power of curses, or the story of Adam and Eve. All of these provided fertile ground for the imagination to flourish, and with their inevitable suppression I think – as with the conquest of the moon, with Neil Armstrong and his great fat boot – that something has been lost.

One day, I found myself wishing that one in particular was still available to us, and that was the idea of spirits. By that I mean disembodied beings, supernatural entities able to speed through the world and appear and disappear at will, some malevolent maybe, some benign, and if you ask me to give you an example, I have one ready to hand: Shakespeare's Ariel, attendant spirit of *The Tempest*.

Ariel, you may remember, is bound to serve Prospero, the magician-duke who has been deprived of his dukedom of Milan by his evil brother and exiled, with his young daughter, to a desert island. Ariel flies hither and thither doing Prospero's bidding – he whips up the storm which brings all the characters together so that the story can be resolved – but he is also desperate for his freedom, which in the end Prospero reluctantly grants him.

Ethereal, insubstantial, even androgynous (I'm saying 'him' for the sake of convenience), unbound by gravity, unburdened by human clay, Ariel is a creation who brings to life the ephemeral longing in us to be lighter than air. But *The Tempest* being what it is, there is more to him than a pet sprite, especially if we see the story of Shakespeare's last play, as most of us do, as auto-biographical: Prospero giving up his magic at the end, is Shakespeare saying farewell to art. In this reading, it is not hard to see the attendant spirit the magician is so reluctant to set free as Shakespeare's own imagination, to which, as old age approaches, he has to say goodbye. His great gift had roamed the world at his bidding, creating storms of his own, and unfor-gettable characters and unforgettable poetry, but now, willy-nilly, he has to say farewell to it and go and be an ordinary citizen – albeit the wealthiest – living in a small market town in Warwickshire and waiting for death (it took four years to come).

That Shakespeare could choose a spirit, a 'tricksy spirit' to represent his own extraordinary, soaring, wandering gift, creating such a dazzling metaphor, was singularly fortunate, and due to the fact that science had not yet consigned such beings to the dustbin of superstition; but to us, such choices are not open. For us, spirits are over and done with, alas, and we cannot compare anything to them; and one spring, I spent several days thinking of this with regret, as I struggled to find some way of expressing my jubilation at seeing, on a sunny Sunday morning in March, the first butterfly of the year.

For it was indeed a brimstone, a bright yellow brimstone. Using science, and rationality, I could have told you quite a lot about it: that it was an arthropod, and among the arthropods, it was an insect; that it belonged to the insect order Lepidoptera, and in that, to the butterfly family Pieridae, the whites; that its scientific name was *Gonepteryx rhamni*; that it had overwintered as an adult, one of only five British butterfly species to do so (the other fifty-three pass the winter variously as eggs or caterpillars or pupae); that in its caterpillar stage it had fed on the leaves of buckthorn or alder buckthorn; and that it had hibernated disguised as a leaf, probably in an ivy clump, until the first warm day woke it up.

But that didn't remotely get it. What I saw electrified me instantly; it was the thrilling sign of the turning year, not just of the warm times coming again but of the great rebirth of everything, the great unstoppable renewal, and the brilliance of its colour seemed to proclaim the magnitude of the change it was signalling. It was like a piece of sunlight that had been loosed from the sun's rays and was free to wander, announcing the spring, and I realised that science, which has now given us so much knowledge about such organisms, did not have any way of conveying its meaning at that moment, at least to me.

For if I say to you, I saw an insect, which is strictly true, what will that tell you? Nothing. The categorisation, which conveys the knowledge, immediately begins to flatten the meaning. But if I say to you, I saw a spirit, which is what it felt like, then at once we are in different territory, we are in the territory of the imagination, and we begin to approach the wonder of the event, and the joy of it: that on a Sunday morning in March, in a mundane suburban street in Surrey, I saw the spirit of the spring.

Part of the allure of the first brimstone, and of the first snow-drops (and of the winter solstice, for that matter), has been that their coming is annually awaited, and the response is accordingly intensified; but there have been one or two isolated or unex-pected events, equally marking the year's rebirth, which have also been exceptional experiences and have produced in me an elation I would readily call joy.

One was to witness mad March hares. For at least five

hundred years, 'mad as a March hare' has been a commonly used simile in English, referring to the excited behaviour of the brown hare in the fields as the breeding season arrives, which – legend has it – is so energised as to seem unhinged. Lewis Carroll reinforced the notion by giving the March hare literary identity in *Alice in Wonderland*, and now it is a character and a concept everyone is familiar with without ever glimpsing the creature in real life. Or hardly ever. The March version of it, I mean.

I had seen many hares and had always been greatly taken with them (and glad of them in a country hardly over-endowed with characterful wild mammals). I think it's partly because we have something to compare them to instantly in our minds, which is the rabbit; we encounter rabbits, and become familiar with them, as young children, long before we ever meet up with their hare cousins, and when we do, the differences are apparent at once: hares are much *bigger*, and we have to readjust the rabbit template squatting in our brains. Hares' towering ears and expandable hind legs seem enormous by comparison, as do their bulging amber eyes, and the body is leaner and rangier than the rabbit's: they're all muscle. Built to run. They seem wilder, too, like adventurers compared to rabbits, which seem like stay-at-homes; yes, I read *Watership Down* just like you did, and briefly thought there might be drama in rabbit society, but ultimately, I know you shouldn't really say this, but don't you think that rabbits are just a teeny bit *boring*? When did any rabbit ever do anything interesting?

Nothing boring about your hare. Not only a dashing wild rover of an animal, but also a hint of the supernatural, with any number of magic legends clustering about the beast, not least that hares were actually witches in disguise – something I first came across when I began to read Walter de la Mare and his children's poems, in my late teens:

In the black furrow of a field
I saw an old witch-hare this night;
And she cocked a lissome ear,
And she eyed the moon so bright,
And she nibbled o' the green;
And I whispered 'Whsst! witch-hare',
Away like a ghostie o'er the field
She fled, and left the moonlight there.

But for all that, and for the great pleasure I took in seeing
hares, throughout my life, I had never actually witnessed the
behaviour that gave rise to the legend, that in March they were
mad – above all, the 'boxing', when they rise on their hind legs
and square up to each other like prizefighters in a ring. I thought
of it sometimes; it felt like a notable gap in my experience. So
when, one year, circumstances arose in which I was offered the
chance to go out with a regular and expert hare-watcher, in
March, I jumped at the chance.

Gill Turner was a friendly woman in her early sixties living
in Hertfordshire, about twenty-five miles north of central London,
and she had been watching, recording, and photographing hares
ever since a chance close encounter with an animal nearly two
decades earlier had sparked her interest. She was devoted to them.
There were hares in her own area, but to show them to me at
their best, she took me another twenty miles north to where
the great arable plains of eastern England were beginning, the
vast hedgeless fields, the 'wheat tundra'. There she had made
friends with a farmer who, unusually, liked his hares too much
to shoot them, and so they were flourishing on his land; but
because there was a significant threat to the animals from men
illegally engaged in coursing – the competitive pursuit of hares
with dogs, usually pairs of greyhounds or lurchers – she was
keen for me to give nothing away about the location. 'It can
be pretty grim,' she said. 'They come from all over the country

to do it. If he [the farmer] calls the police, they [the coursers] dump the dead hares on his doorstep.'

So no precise details about where. But the landscape was fascinating, low rolling hills of thin topsoil, seeming very bare, with scarcely a windbreak: an eastern England archetype. 'My God, it's cold here in the winter,' Gill said. It was the morning of 2 March. It was cold and dry. To my delight, there were lapwings calling and displaying. We walked down a path through a small wood and out on to the plains, and at first we saw no sign at all of *Lepus europaeus*, while I asked Gill about the animal's attraction for her.

People were misinformed about it, she said, mentioning the female hare's habit of giving birth to her young, the leverets, in the open field, in a mere depression in the grass called a form, in contrast to the rabbit's comparatively safe birthing chamber in its burrow. 'People said hares are poor mothers, that they have their babies on the ground and leave them. But when I took time to study them, I found they were brilliant mothers, absolutely wonderful mothers. Before they have their young, they spend weeks watching who goes across their land. Female hares will have their young almost in the same spot all their lives; they will find an area that's very safe, and before they even mate, they will know what crosses that area.'

Over the years she had built up a rich body of observation, with many small curious details. Hares took dust baths, she said. 'When they find an area of dry powdery sandy soil, they will go and roll in it. There's a sort of etiquette. A hare will wait for another hare to finish before that one goes in.' The youngsters would gather together in a group and chase other creatures. 'I've watched them chase crows, pheasants, anything that happens to settle near them. They'll chase them away.' You could tell a young hare, she said. 'The snout is shorter. There's no damage to the ears. Older hares have damage to the ears, especially the bucks.' And of course, she had often witnessed the mating behaviour, the mad March hares, the chasing and the boxing, which was once thought to be two males battling over a female, but is now thought to be nearly always a doe hare fighting off an unwanted buck's advances.

Hares gradually began to appear on the fields as we walked deeper into the farmland, usually fairly distant, the odd animal here and there, and then small groups of them, scattered around, some of them closer. Once you got your eye in they were conspicuous, not least for their ears, black-tipped and perpendicular, and it became clear they were plentiful. God bless the farmer. They seemed to be calmly going about their business – 'nibbling o' the green' – but as I scanned one of these groups two animals suddenly leapt at each other face to face with a flurry of forepaws and I cried out: 'Yes!' Gill smiled and said: 'There you are.' It was just for a second, though. Quick as a shooting star. I found myself wondering, had I really seen it? And then it happened again, this time for longer, and in the binoculars I could see the hares' white bellies as they danced at each other, circling around upright on their hind legs with their front paws frantically flailing, trying to land blows. They stopped and sat watching each other for a short while a few yards apart, like boxers in their corners, then – almost as if the bell for the next round had rung – came together again in a collision that was truly aerial: each leapt up towards the other and their whirling paws clashed while they were still in mid-air, before the frenzied sparring continued on the dancing hind legs, and in my mind I heard the shout of young schoolboys alerting their companions when a scrap breaks out in the playground: *Fight! Fight! Fight!*

Up and down the rolling hills we saw it then, in different groups of the animals, short outbursts of boxing, longer matches, interspersed with frantic chasing of one hare by another, which more hares would sometimes join – Gill said it was the youngsters, who would do it even though they didn't know why they were doing it – and watching it all, I found an unstoppable elation spreading through me, which was more than just the excitement (though it was tremendously exciting) and more than just the gratification of finally catching up with the rarely seen reality behind a figure of speech. There was a sense of privilege: I was seeing a part of the reawakening, of the movement towards new life, which was extraordinary, which you wouldn't ever normally see, and that was what was joyous.

It was like seeing the sap rising.

It was like seeing the sap rising at supersonic speed.

The other unconventional marker of the reviving year which I have experienced and found joy in is so unusual that I don't actually know how to characterise it, as it emerges from modern electronics.

In the summer of 2011 the British Trust for Ornithology, Britain's leading bird research organisation, began a project in which I had a strong personal interest. It concerned the cuckoo, the European cuckoo, the bird with a double claim to fame: it lays its eggs in other birds' nests, and its two-note *cuck-coo* call when it arrives in Britain in April is the best-loved, most notable, and most distinctive of all our sounds of spring, being a perfect musical interval (a descending minor third).

I was interested in what the BTO were doing because two years earlier I had written a book about the British birds which are summer visitors, the migrants from sub-Saharan Africa, such as the swallow, the nightingale, the willow warbler, and the cuckoo in particular – the spring-bringers, I called them – some of which, the cuckoo included, were undergoing alarming declines in numbers. It was difficult to know where the problem lay, as migratory birds 'live in multiple jeopardy' – they may face difficulties on their breeding grounds in Britain, or on their wintering grounds in Africa, or on the immense and gruelling journeys between the two which they annually undertake.

Quite a lot had by now been established about what cuckoos did during their summer breeding season – how they outwitted the other birds in whose nests they laid their deceiving egg, such as reed warblers or meadow pipits, how the cuckoo chick got rid of its rival chicks once it was hatched, to monopolise the attention of its foster-parents – so the BTO research project was an attempt to focus on the rest of the cuckoo's year, the journey back to Africa and the time spent there, to see if that might offer any clues as to its decline. Virtually nothing was known of it. There was a single piece of relevant data: a cuckoo ringed as a chick in a pied wagtail's nest in Eton in Berkshire, in June 1928, was found dead in Cameroon in West Africa in January 1930.

That was it.

The rest was a blank. Where do cuckoos from Britain go in winter? Nobody had any idea.

The project aimed to remedy this by the use of modern communications: the miniaturisation of satellite transmitters had now gone so far that they could be fitted to birds and the birds' progress followed step by step around the globe. It had already been done with larger species such as ospreys, and by 2011 the satellite 'tags' were small and light enough for a cuckoo to carry one without being hindered in its flight. In the event, five male cuckoos were caught that May, all in East Anglia not far from the BTO's headquarters at Thetford in Norfolk, and ringed and fitted with satellite tags before being released.

In a clever move, the BTO gave them names. In the past, subjects of such serious and expensive scientific research (the tags cost £3,000 each) would probably have been labelled XPWS137 to XPWS141 or some such, but the trust had a sharp eye for public support and it named the five Clement, Martin,

Lyster, Kasper, and Chris. They sounded like the members of a boy band. And it went a step further in sassy modern media terms: it gave them each a blog, on which details of their separate journeys would be recorded, and which could be followed on the BTO website by anyone, more or less in real time.

The project paid off instantly and spectacularly, demolishing once and for all the accuracy of the ancient cuckoo nursery rhyme:

> In April
> Come he will;
> In May
> He's here to stay;
> In June
> He changes his tune;
> In July
> He prepares to fly;
> In August
> Away he must.

August, huh? Well, Clement left Britain for Africa on 3 June and was in Algeria by 13 July. You start on your winter and it's not even midsummer yet? The BTO scientists were astounded. Clement hadn't changed his tune in June. He'd simply scarpered, and he was soon followed by Martin, Kasper, and Chris (although Lyster stayed in the Norfolk Broads till mid July). That was only the start of the revelations. The researchers were further taken aback by the direction and nature of the migratory journeys, as they unfolded, for they split into two vastly distant routes but ended up in the same place. Three of the birds, Chris, Martin, and Kasper, flew down through Italy, over the Mediterranean, and straight across the Sahara desert, while the other two, Clement and Lyster, went to Spain and flew around Africa's Atlantic edge, more than a thousand miles to the west. Yet by the end of the year they had all recongregated in the same, little-known part of the continent, the Congo river basin. Cuckoos from south-east England, it was revealed, fly 4,000 miles to Congo for their winter (not the Democratic Republic of the Congo, the DRC, the huge former Belgian colony, but its smaller neighbour to the north, Congo-Brazzaville, the French Congo that was). Nobody knew that. No one had any idea. It had been assumed that they probably went to West Africa, to Senegal or somewhere. Even more surprising was just how close to each other they ended up. By the new year, Clement, Martin, and Lyster were all wintering on the Téké plateau north of Brazzaville, a sparsely inhabited area of grasslands with forests along the rivers, Kasper was on the Téké plateau's southern end, while Chris was further to the north-east, just over the border in the DRC.

As someone with what you might call a professional interest in cuckoos, I was wholly absorbed by all of this, and followed the fortunes of the five birds closely from the start: you could see precisely where they were (or at least, where their tags had last transmitted) at any given time, on Google Earth. It was wonderful, cutting-edge ornithology, it was thrilling to see the discoveries as they happened, to watch the ancient migration mysteries unfold, surprise after surprise. But a greater surprise lay in store.

On 7 February 2012 I logged on to the BTO website and its cuckoo pages and read the summary of what was happening with the birds, which had now been in Congo for two months. There were no recent data on Chris or Clement. Lyster had moved 75 miles north to Ndzakou. Martin had moved 90 miles north and was close to the Likouala river. And Kasper had leapfrogged these two, and from further south, had moved 350 miles north to be close to the Congo border with Gabon.

Something stirred in my brain.

I read it again. Lyster had moved north. Martin had moved north. Kasper had moved north.

I clicked on the map and looked at the movements, the thin straight lines, orange for Lyster, green for Martin, and yellow for Kasper.

They were all pointing in the same direction. Northwards. Northwards towards . . . where I was sitting . . . and with a mixture of astonishment and intense delight, I began to realise what I was witnessing, on the screen in front of me.

They were coming back.

The great migration cycle had begun again, and it suddenly dawned on me, however impossibly hyperbolic it might sound, that I was seeing something no one in human history had ever seen before.

I was watching the spring coming, from 4,000 miles away.

I wanted to shout out, at the top of my voice. I wanted to run out into the street, grab the first passer-by, drag them in to my computer screen and cry, *Look! Look! It might be February, it might be freezing, but here comes the spring! Down in central Africa! On its way to us! Right now!* But the pathetically conformist part of me prevailed and I merely sat there, awestruck at what I was watching, and let the joy wash over me (for joy it truly was) while I wondered what had triggered the great shift. What was the cue? Some whisper in the tissues of faraway Norfolk, and its reed warblers, and their tempting riverside nests? A change in the African rainfall pattern? A variation in day length? Whatever it was, it had instructed the birds, in imperious terms which brooked no denying: *start again.*

From then on, of course, I closely followed their return journeys, which provided another revelation – all British-breeding cuckoos, whichever route they take to fly south for the winter, fly back north in just the one way, with a major detour, a major left turn, to the West African rainforest; they swing across to Nigeria, Togo, and Ghana, where the spring rains bring forth a burst of insect life, which the cuckoos use to refuel before the arduous crossing of the Sahara.

They need it. For it was a period of cuckoo tragedy as well as cuckoo triumph, illustrating just how dangerous and demanding the annual migratory treks can be. Clement, who had left England so astonishingly early, died in Cameroon on 25 February, cause unknown – he could have been taken by a predator or shot for the pot by human hunters – while Martin died, the scientists believed, in an unseasonal and severe hailstorm near Lorca in southern Spain on 6 April, and Kasper stopped transmitting in Algeria on 9 April, although it was thought that could be a case of tag failure. But Chris and Lyster successfully made it to England in late April – and on 30 April the BTO team actually went and found Lyster in the Norfolk Broads, and caught sight of him, to say welcome back. They were elated.

At one remove, I was elated too. Not only for the happy return, but to have witnessed the detailed unfolding of these 8,000-mile odysseys, through some of the most starkly differing landscapes on the planet. Since leaving the placidity of East Anglia, the cuckoos had plunged into extremes: they had crossed the world's biggest desert, the Sahara, and the world's densest rainforests, in West Africa. They had flown around the Atlas mountains and the western Congolian swamp forests (legendary home of the mokele-mbembe, Africa's version of the Loch Ness Monster). They had seen not only France, Italy, Spain, and the Mediterranean, but also Mali, Niger, and the Central African Republic. They may have seen Paris; they may have seen Timbuktu.

This sense of wild creatures wandering at will through the world, in a way mere humans never can, was captured by Ted Hughes in a poem called 'October Salmon', where Hughes looks at a dying salmon which has come home to spawn and expire in its Devon river after its journey to the seas off Greenland. 'So briefly he roamed the gallery of marvels!' Hughes writes, and indeed, the five BTO cuckoos had roamed their own gallery of marvels, with the remarkable advantage, to us, that we could follow them doing so (and the project continues today).

I found all of it inspirational. I watched their wanderings with amazement. But nothing in it, nothing, compared to the moment when I suddenly saw that they were returning, the bearers of the two-note call, the perfect musical interval, which would ring out over the English countryside and proclaim indisputably that the new season was here; when I saw that the great eternal cycle had started again. I know it was the most extraordinary signal of the reawakening world that I will ever witness, and the sheer joy of it remains: it was like the joy of the winter solstice, the first snowdrops, and the first brimstone, all rolled into one, that day in February when, sitting at my computer screen, I saw the spring coming, four thousand miles away in the heart of central Africa.

There is one more marker of the reawakening earth which has given me joy, and that is blossom. It is a curious and charming peculiarity of English that it has a special word for the flowers of trees – other languages don't, they simply call them tree-flowers, they say *les arbres en fleurs* or *die Baumblüte* – and this echoes the inchoate feeling I had for years that there was something in the nature of blossom which was special in itself. If I saw a blossoming cherry tree in a bed of flowering daffodils, eye-catching though both might be, I would be more animated by the white blossoms above than by the yellow flowers beneath.

Yes. I would.

Why should that be?

I used to think it was because blossom, especially on fruit trees such as apples, cherries, and plums, tends to appear in clusters, globular and luxurious, like the clumps of fruit they will turn into, and in their lush opulence these seem to be the very essence of the floral. But now I think the attraction is simpler and deeper, it is a temporal one: there are flowers all the year round, but there are flowers on trees, generally speaking, only in springtime.

So blossom is of its very nature a banner, a bright banner with spring written on it, waving in the wind, and over the years I have developed in my mind a blossom calendar of my own, with its own special occasions eagerly awaited. You can kick off such an almanac right at the start of the year with some rarefied species such as winter-flowering cherry, but for me it begins in early March, with a foreign one, transplanted to England: magnolia, the tree of enormous, blowsy blooms. In my corner of the world, suburban west London, magnolias flourish in front gardens; on my commuter's morning trudge from door to station I used to pass several of them, and as February ran its course it was impossible to ignore their great buds, erect, fleshy and tumescent, swelling until they were as fat as light bulbs. It was like watching fizzing fireworks ready to explode, until eventually, *Bang!*, they did, and suddenly before somebody's front window was a smooth bare tree covered in waterlilies.

Magnolia bud-burst is pretty regular, and for years I noted the date for a spectacular white example at the bottom of my road, now sadly no more: it averaged 9 March, and it always brought forth a punching of the air from me, a *Yes!* like the first glimpse of the boxing hares. Joy in the first flowering tree. The world was unquestionably turning. It was a stunning as well as an uplifting sight, for whether white or cream or pink or yellow, magnolia blossoms are exuberant and tropical things, seeming far too exotic to be English, and of course they aren't: their natural home is split between two far more glamorous parts of the world, in botanical terms, Asia and Central America (including the southern United States). But the intrepid efforts of plant collectors over the last two centuries have brought many of the two-hundred-odd species back to Britain, where, especially in London (and in Kew Gardens in particular), they have flourished.

Their lavish flowering at what is usually a chilly and inhospitable time, while the surrounding trees are wondering hesitantly whether to put out a leaf or two, is one of the reasons why magnolias are so eye-catching here, such a source of pleasure. Another is that in the urban context, where I tend to see them, their bold brightness shows them off particularly well against brick or stucco. But perhaps the reason why they are special anywhere is the structure of the blossoms, which are not only whacking great things – like white doves nesting in the tree, a keen gardener friend once said to me – but appear uncomplicated: like the tree itself, they have simple, clean lines. In style terms, they're minimalist.

This is no doubt because magnolias are among the most ancient and primitive of all the flowering plants (like the water-lilies their blossoms superficially resemble); they give us a hint of what flowering plants might have looked like when they first developed from conifers around one hundred and fifty million years ago. If you look at a magnolia bud, it looks very like a closed pine cone; you might say a magnolia flower is what an open pine cone became, once it had evolved colour and nectar to attract the winged insects to pollinate it, which were evolving at the same time . . . yet I am cheating, really, writing about this. I am concerned with the natural world, and this is verging on the horticultural. As magnolia is non-native, you will struggle to find it in much of Britain (although there are four national magnolia collections); it's just that it's been such a significant part of my own experience of spring blossom, I was reluctant to leave it out.

The next item on my blossom calendar, however, is spread across the country, and this is blackthorn. A member of the Prunus family, the stone fruit – the plums, the cherries, the peaches, and the almonds – its scientific name is *Prunus spinosa*, or the thorny plum, and it produces sloes, those small black plums which are mouth-puckeringly astringent until the frosts get at them in October. Then they sweeten and can be used to make sloe gin, one of Britain's great native drinks, out-topping even the fruit *eaux de vie* of France. Sensational stuff. Don't get me started on it. Another benefit of the blackthorn is its wood, which makes prized walking-sticks – in Ireland it was traditionally used to make the shillelagh, the fighting club – and a third is that its leaves provide the larval food for two of Britain's less common butterflies, the black and brown hairstreaks. The black hairstreak's a bit dull, to be honest, but the female brown hairstreak is one of our loveliest insects, with glowing golden bands across her brown forewings, and you can only see her when she descends from the treetops in late August and September to lay her eggs on blackthorn twigs. One of my most prized possessions is a painting of a female brown hairstreak next to a bunch of ripening sloes by Richard Lewington, insect artist supreme: autumn glory, I think when I look at it.

But spring glory, even more, is what the blackthorn furnishes. When the bush flowers, usually in mid to late March, it looks as if it is covered in hoar-frost rather than weighed down with fat hanging blossoms in cherry-tree style; it looks like trees do on those mid-winter mornings when you wake after a night of freezing fog and every black branch seems to have been dusted with sugar. The reason is that the flowers on the blackthorn appear before the leaves, so the whole arrangement is more spindly and delicate, skinny bare branches that seem to have been sprayed with white. Skinny or not, they transform the landscape. Blackthorn hedges are widely planted and in the monochrome countryside of March and early April they provide the first substantial burst of colour: a month before the greening comes, there is a whitening of the world. I once drove in early spring from Brighton to London, and the A23 was bordered with blossoming blackthorn for mile after mile: every few yards, for ten miles, twenty miles, thirty miles through the Sussex countryside, there seemed to be a blackthorn bush dressed from head to foot in white, and I wondered how many of the drivers pelting along the dual carriageway were appreciating its spectacularly ornamented flanks. Eventually I found a blackthorn-fringed lay-by, pulled in, and greedily broke off two of the blossoming twigs and drank deep of their honey scent. They travelled with me on the dashboard all the way home, thorn branches frosted in numberless small white petals. I loved them. I love them every year.

After the blackthorn, in April, the blossom marked in my calendar comes thick and fast. In the small garden of our house we are blessed with an apple tree (a Bramley seedling), a true cherry, and a lilac, and in most years we tend to have a few days when all three are out together, decorating the garden extravagantly in pink and white, pure white, and pale lavender-blue. At such times, drawing back the curtain of my daughter's bedroom window causes a sharp intake of breath as the apple blossom is right outside and fills the whole window pane; while in the streets around us, the horse chestnuts, their new leaves an iridescent emerald, top off their transitory magnificence with the biggest blossoms of all, the white roman candles, bulky as pineapples. And then one more before the calendar closes: the hawthorn, or the May blossom, named after its month, rich and luxuriant in its hedgerows – cream to the blackthorn's sugar.

All of these are beautiful, but it is not just their beauty which so strongly affects me, it is that they are markers of the turning year: the very act of setting eyes on blossom locates you in the springtime, and I think our bond with nature is very obvious in the power of the natural calendar and its events to move us to joy, in the fact that the annual rebirth of the natural world is not a matter of indifference; or at least, it is not to me, and I know it is not for many people. On occasions, it has moved me to a joy so intense that I have been at a loss as to how to respond.

One such experience took place in France. For ten years, my wife and I and our two children spent many of our holidays in an old farmhouse in southern Normandy, in the rolling, wooded hills of the Perche, the ancient medieval county which is home to the Percheron great horse, and which is bypassed by most British tourists. One of the house's attractions was its large garden, which held many songbirds with spotted flycatchers the most thrilling visitor, as well as constant swooping swallows and linnets and yellowhammers singing on the telephone wire, plus the odd mammalian surprise: red squirrels came in from time to time from the wood across the road, and once my wife saw the snaking shape of *la fouine*, the stone marten. But for me perhaps the most significant attraction was a profusion of insects now but a distant memory in insect-impoverished Britain: the butterflies were splendid, from swallowtails to fritillaries, and at night the moths were magnificent, as I know since I took my moth trap there, unashamed nerd that I am. Among a great moth menagerie there were Jersey tigers and crimson under-wings and several different species of hawk moth, not least the privet hawk moth – big as a bomber, or so it seemed when I first set astonished eyes upon it – as well as the angle shades and the buff arches and the setaceous Hebrew character and the large yellow underwing and all sorts of other stuff, and there was much more than Lepidoptera. Sometimes we would be visited by the bulkiest of all the bees, the violet carpenter bee, navy blue and the size of a cocktail sausage; and when dusk arrived, the children would be entranced by the tiny points of luminous green radiance in the grass, as the female glow-worms lit up their lamps for passing males.

The rear half of the garden was given over to a small old orchard of fourteen different fruit trees, with apples and cher-ries and peaches and several varieties of plum, including damson, *quetsche* in French, and greengage, which is *reine-claude*, and *mira-belle*, which is the same in both languages and which, if you get it at the right moment, is the acme of all fruits. It is small and round and greenish-yellow as it slowly ripens and then it tastes perfectly pleasant, with what one might call a generalised plum taste; but right at the end of its ripening, in the day or two before it drops off the tree, its skin colour deepens to old gold with red spots and then, ah then, its taste is like nothing you have experienced, the most subtle sideways variation on sweetness your palate will ever be blessed with.

But the orchard held other blessings. In springtime the blossom was spectacular, especially on a couple of the pure white cherries, which seemed, like the Easter trees in A. E. Housman's matchless lyric ('Loveliest of trees, the cherry now') to be 'hung with snow'. And there was yet a further blossoming too, and that was the birdsong.

For some years I have thought of spring birdsong as blossom in sound. This takes us I suppose into the realms of synaesthesia, the interpreting or experiencing of one sense via another, not a concept I have ever found particularly rewarding or fruitful, despite its promotion by numerous prominent figures in the arts; but once the suggestion flowed into my mind, listening to willow warblers singing on Skye, their silvery falling cadence softening the severity of the northern landscape as much as flowering trees might do, it would not leave. In the orchard in France the birdsong was fulsome: we woke every morning to a chorus of blackbirds and song thrushes, robins, wrens, and chaffinches, and best of all a blackcap, with the most mellifluous, melodious song you can imagine, and I began to think of this as blossom, as much as the blossom was blossom; and then, in the most extraordinary experience – at least, it was for me – they merged into one.

For one late April the blackcap was singing unseen, deep in a hedge, and it was joy-inspiring; and across the garden was the most gloriously flowering of the cherry trees, and that was joy-inspiring too. Then on a Sunday morning – I remember it precisely – the bird moved into the tree and began its song.

I was struck dumb in amazement.

Here was this God-given, blossoming snow-white tree, which was breathtaking in its beauty; and here was this God-given, breathtaking sound coming out of it. This tree, this tree of trees, was not just an astonishing apotheosis of floral beauty. It now appeared to be singing.

The rational part of me couldn't cope. It was all too much, and it fell to bits. I had gone way past simple admiration into some unknown part of the spectrum of the senses, and there was only one possible response: I burst out laughing. And there, in the exquisite fullness of the springtime, was the joy of it.

6

Joy in the Beauty of the Earth

If we start with the calendar, and the markers of the world's reawakening, we can go deeper into joy with the beauty our world possesses. I will single out two examples which have brought joy to my own life, one of colour and one of form.

I have never seen it actually remarked upon, but it is clear that the earth did not have to be beautiful for humans to evolve; we could have had a planet which perfectly well sustained us with air and water and food and shelter, without offering us aspects of itself which also lift the spirits and catch at the heart. For example, for a substantial part of the time that life has existed, the land surfaces of the earth were very likely to have been just one colour, the colour of the plants which from about 450 million years ago began to cover the ground, gradually becoming taller and forming forests. They were green, and so the earth, almost certainly, was green too. Many shades of green, perhaps. But green. For maybe 300 million years, give or take the odd epoch. Yet then the time came when some plants began to use insects instead of the wind to move their pollen around, and evolved reproductive organs with brightly coloured petals to advertise their presence and catch the insects' eyes, just as the magnolia did – and in a great outburst of beauty, flowers were born, and exploded in size and shape and colour and number. While the ancient seed plants without flowers, like the conifers and the cycads, now total only about a thousand species across the world, there are more than three hundred and fifty thousand plants whose reproductive systems are floral.

The emergence of flowering plants was one of the great revolutions of life on earth, but it didn't have to happen, and certainly, nothing said it had to happen before we came along: we might well be living happily – in so far as we can live happily at all – in an all-green world still, and perhaps we would not miss what we had never had. As it is, most of us take the existence of flowers wholly for granted, save for the occasional perceptive soul such as the novelist Iris Murdoch, who had one of her characters say (in *A Fairly Honourable Defeat*, in 1970): 'People from a planet without flowers would think we must be mad with joy the whole time to have such things about us.'

They very well might. It is a peculiar property of the earth that it offers us beauty as well as the means to survive, but it is also a wondrous property, and it greatly moves us – as behaviourally modern humans, anyway. Hence over about forty thousand years we have steadily formalised our appreciation and our celebration of it, in what we have come to call art, from Lascaux to Leonardo. Until, that is, the last century. In the last hundred years or so, with the advent of modernism, a new artistic philosophy for an industrial age (and also for a world whose optimism had been irreparably fractured by the First World War), many of our society's high cultural elites have consciously rejected the primacy of beauty, seeing its veneration as outmoded and complacent, and holding that the true purpose of art should be to challenge preconceptions; and they have largely forgotten all about, or simply ignored, where beauty comes from in the first place, which is the natural world.

In more recent decades the process has gone even further,

and beauty has become *suspect*. I spent the middle years of my life watching a novel notion take shape in my culture, and gather force: the undermining of the idea of excellence. My early years were lived in a world where the worth of excellence seemed to be taken for granted by everybody across the political spectrum: it was a cornerstone of the post-war meritocracy, as indeed it had been a cornerstone of European civilisation since classical Greece. But things have changed. In the last quarter of a century two opposing political visions have prevailed in two different sectors of our society. In economics, the vision of the free market, the vision of the Right, has carried all before it; while in social policy, the key idea of the Left, that of egalitarianism, has won no less signal a victory. Yet this is not egalitarianism as in equality of opportunity, a political concept at least as old as the American Declaration of Independence in 1776; this is a new egalitarianism, as in equality of outcome. The key idea is that there should be no more losers – not hard to sympathise with that – but therefore the corollary is that there should be no more winners, either. In anything. No more excellence. No more elites. So a concept, say, which has been central to European poetry since the troubadours of Provence began it all in the eleventh century, the praise of feminine beauty, more or less ceases to be valid, because it is seen as offensive to women who may not be thought beautiful, or it is seen as patronising to women who may have many gifts other than accidental beauty. If not invalid, at the very least such praise becomes questionable. Quite suddenly. Just like that. Petrarch should try singing the praises of his Laura today, and her beautiful eyes, and see if he gets published.

I am not making a stand against this development. I am not even suggesting it is wrong, or bad. I am merely saying that it has undeniably happened, and it is noteworthy. Beauty has in some quarters become bound up in ideology, it has become associated with privilege, it is seen as the plaything of those who have greater advantages, and I have found myself wondering (only in idle moments, of course) if the day might not come when to express open and unqualified admiration for an orchid, say – I mean for its beauty, its elegance and its glamour, all qualities many orchids undeniably possess – might be thought inappropriate . . .

Probably not. But there is no denying that the veneration of the beauty of nature, which Wordsworth made the fount of his philosophy, has largely ceased to figure in high culture since modernism contemptuously swept it aside; and modernism's triumph was of course comprehensive, in painting and sculpture, in music and in poetry. In the early part of the twentieth century, for example, there was a substantial group of English poets collectively known as the Georgians who wrote extensively about nature and were read by large audiences; some were quite good, some were not, but all except one were consigned to lasting oblivion by T. S. Eliot's *The Waste Land* in 1922 and the modernist revolution which followed (the exception, of course, being the wonderful Edward Thomas, who was anyway very much more than a 'Georgian nature poet'). We retain the legacy of those attitudes. So beauty in general and the natural beauty of the earth in particular have gone largely unsanctioned as objects of relevance by the cultural elites of the twentieth and now of the twenty-first century, and we hear little of them from those quarters; and yet, of course, many ordinary people who do not feel they must be aligned with prevailing cultural modes of thought have been drawn to the beauty of nature as much as people ever were, and I am one of them. Let me tell you about a wood. Five times in the one week, I went to this wood. Five separate trips, on five successive days. And each time, after the first time, I stopped at the gate, I paused before entering. I savoured the moment. It felt like the minute before sex, with a new lover who is making ready – the elevated heartbeat, the skin-prickle, the certainty of impending pleasure – but it was even more than that, it was the anticipation of a sort of ecstasy, at beholding what the wood contained, hidden in its depths, which was something truly exceptional, as exceptional as a crashed flying saucer, I found myself thinking . . . Each time I stopped at the gate I said to myself, *I know what is in there* . . .

It was a blue.

It was a blue that shocked you.

It was a blue that made you giddy.

It was a blue that flowed like smoke over the woodland floor, so that the trees appeared to be rising out of it, a blue which was not solid like a blue door might be solid but constantly morphing in tone with the light and the shade, now lilac, now cobalt, a blue which was gentle but formidably strong, so intense as to be mesmerising: at some moments it was hard to believe it was composed of flowers. But that was the beauty and the joy of the bluebells, their floral richness and their profusion, a dozen blue bell-heads nodding on every stem, a hundred thousand stems pressing together in every glade until it ceased to be plants, it was just an overwhelming incredible blueness at the bottom of a wood.

They make a remarkable phenomenon, bluebell woods: you enter and are amazed. They are one of the specialities of the natural world in Britain – the home of the flower is the damp Atlantic fringe of Europe and we have more of them than anywhere else – but of course they are not just a speciality, they are a glory, one of the two supremely beautiful habitats of my native land. In a countryside razed by the farmers in the Great Thinning, denuded of its former wealth of living things, they offer a miraculous, and perhaps the most magnificent, survival of abundance; the very profusion of the flowers, packed tightly together over wide areas, and always in descriptions prompting words such as *sheets, carpets, swathes*, is a major part of the attraction. But for me, it is not the key part, for the same effect can sometimes be seen with sheets of snowdrops, or with carpets of wood anemones, or with swathes of ramsons (wild garlic), all of which are stunning yet would not draw me back, I am sure, for five days in a row. The key attraction is something else. It is the blue.

When people interested in aesthetics discuss beauty nowadays, and try to get to the heart of it, it seems to me they seldom stress colour; they tend to put much more emphasis on harmony, as in harmony of proportion, certainly in areas such as architecture or the human form. I fully see the force of this and would in no way dissent from it; I would only say that for me personally, colour in nature has an allure which is compelling, and the more striking the hue, the more exceptional and wondrous a place the natural world seems to be. Sometimes this can be very simple, as with the large copper butterfly, which died out in Britain in the nineteenth century, was reintroduced to Woodwalton Fen in Cambridgeshire in the twentieth, and sadly, died out again; but you can see it (and I have) in continental Europe. The large copper male bears four wings of the most lustrous bright orange. Nothing intricate: simply that. The purest, most saturated orange you can possibly conceive of; indeed, it may be more than you can conceive of, which is perhaps the source of the delight when you eventually set eyes on it, as your sense of what the world can contain is suddenly enlarged. In fact, nature's ability to generate colours and colour combinations you have never seen before is endless, and that is part of the thrill, part of the joy of the beauty of the earth. If, as seems likely, for 300 million years the land was just one tone, green, then look at what we have now: there are those 350,000 and more species of wild flowers, as well as 200,000 butterflies and moths with painted wings and above a million other insects, 10,000 birds, 10,000 reptiles and 7,000 amphibians, virtually all of them using colour to differentiate themselves, not to mention perhaps 8,000 species of brilliantly bright coral reef fish. Could we ever list all their colours?

We would start, of course, with the eleven basic colour terms in English (ordered hierarchically according to the Berlin-Kay hypothesis): the black and the white; the red, the yellow, the green and the blue; the brown, the purple, the pink, the orange, and the grey. But that doesn't remotely get it. What about the scarlet, the russet, the violet and the olive? The crimson, the sulphur, the indigo, and the emerald? What about the magenta and the turquoise, the ivory and the aquamarine, the lavender and the maroon, the coral and the mauve? As the gradations become more refined, so they seem to stretch in a line into a misty distance of subtlety – terracotta, lime, amethyst, fawn, jasmine, tawny, amber, cerise, butterscotch, mahogany, teal, beige, oyster, cerulean, ox-blood, fulvous, vermillion, tourmaline, gamboge – not just glowing singly but in heart-stopping combinations, in their patterns bold and patterns delicate, in their stripes and their spots and their cross-hatching . . . All of them, and so many more which we do not even have names for, are there in the natural world. Colour is its ultimate abundance.

They are there, of course, for a reason: they are functional. They have evolved through Darwinian natural selection purely to enhance survivability in their host organisms in a whole series of ways, from making them conspicuous to making them blend into the background, from making them frightening to predators to making them desirable to potential mates, from making them appear fit and dominant to making them appear poisonous . . . Yet for us this, the instrumental side of it all, which is fascinating, had to be uncovered by science, by evolutionary biology. It is not what we humans, possessed of an aesthetic sense, instinctively take in when we look at a creature like the Jersey tiger moth; we do not see that its black and cream striped forewings are camouflage to break up its outline and its blue-spotted crimson underwings are there to be flashed in the startled face of a predator to give the moth an extra millisecond to get away: we just see that it is gorgeous. And so with flowers, and butterflies, and birds, and other organisms without number: they have their colour functionality; we have our joy in them.

It is part of our great fortune in finding ourselves on a planet that did not have to be beautiful for us to evolve on it, but turned out to be beautiful beyond what, in a monochrome world, we might ever imagine. Let me take just one example of a group of organisms I do not think we could invent, in their sheer colour diversity: the wood-warblers of North America. Although unrelated to the warblers of the Old World, such as our chiffchaff and willow warbler, these have co-evolved to occupy a similar ecological niche, as small insect gleaners of the treetops; yet whereas our birds are by and large plain-looking, dun creatures, mainly brown and olive-green – they do their signalling by song rather than appearance – the fifty or so warbler species of America display a range of flamboyant colour and patterning which is quite unparalleled (at least, with the male birds in their springtime breeding plumage). It can often be seen as variations on a theme, such as a black throat with this or a striped back with that, and the colours of this and that are intense – rufous, gold, sky-blue, dove-grey, flaming orange, navy, chestnut – juxtaposed in plumage arrangements which are often startling and make an incredible feast for the eyes. After my first, astonished experience of them, a few years ago, I asked a leading American ornithologist, Greg Butcher, then director of conservation for the National Audubon Society, and a plumage expert, how even natural selection could have produced such scarcely conceivable variety. He said: 'Well, first there was selection for colour; then there was selection for difference; and then the palette was allowed to wander' – which I thought was a charming idea. And how that palette has wandered! To give a single detailed example, the magnolia warbler, the breeding springtime male, has a grey crown, a white eyebrow, black cheeks, and a yellow throat – that's just his head – and then a black back, white wing-patches and a yellow belly marked with thick black stripes. And he's far from the most spectacular: witness the golden-winged warbler, or the prothonotary warbler, which is even more golden, or the black-throated blue warbler, or especially the blackburnian warbler, which below its wings of black and white has underparts of such powerful, passionate orange that American birders have nicknamed it the firethroat. They are fragments of a rainbow, these birds, pieces of a painting, they seem like elements of a great meta-species; their coming, when they migrate from their wintering grounds in Central and South America to breed in the boreal forests of the USA and Canada, represents what is most exceptional, of all that is exceptional, about America's spring, and I have thrilled to see them, even in New York's Central Park, where I once watched an American redstart, one of the most exquisite of them all, a giant black and orange butterfly of a bird, hovering round the trees a few yards from the tourist-thronged Strawberry Fields memorial to John Lennon, shot dead outside the Dakota Building across the road.

No, we could not have invented America's warblers: it is nature we need, to come up with such endless diversity of tint and tone. And yet, breathtaking though it is, I think I am drawn most of all to the intense single colours the natural world sometimes offers, such as the large copper's saturated orange or the pure white of some waterbirds such as egrets, fresh snow against the background of a greeny-brown marsh, or the lipstick scarlet of poppies, or the purple flash along the flanks of rainbow trout, or, indeed, the blue of the bluebells in the wood to which, that springtime (it wasn't very long ago), I returned for five days in succession.

It was the blue which drew me back. I know I am more drawn to blue than to any other colour. Let me give some examples, leaving bluebells for the moment to one side. Two other blue flowers also move me greatly: one dark blue, one pale.

The dark blue is the cornflower, prominent among the plants exterminated from the countryside by Farmer Giles with his unkickable herbicide habit. I have seen far more of them in Normandy than I ever have in England; in France *les bleuets* are held in special affection, as they are the flowers associated with the French soldiers who fought in the trenches in the First World War, the *Poilus*, just as in Britain the scarlet poppies remain the symbol of their English equivalents, the Tommies. The particular aspect of cornflowers which attracts me is that they seem to glow, such is the depth of their colouring – it is indigo, really – but to glow with dark rather than with light, almost as if they are throbbing with darkness and shedding it; and when I got to know them, in mid life, this quality suddenly triggered in me the memory of a forgotten poem I had read and loved as a teenager, which actually articulates the precise idea, hypnotically, almost as an incantation. It is D. H. Lawrence's 'Bavarian Gentians':

> Bavarian gentians, big and dark, only dark
> darkening the daytime torchlike with the smoking
>   blueness of Pluto's gloom,
> ribbed and torchlike, with their blaze of darkness spread
>   blue
> down flattening into points, flattened under the sweep of
>   white day
> torch-flower of the blue-smoking darkness, Pluto's dark-
>   blue daze,
> black lamps from the halls of Dis, burning dark blue,
> giving off darkness, blue darkness, as Demeter's pale
>   lamps give off light,
> lead me then, lead me the way.

Lawrence asking the flower to lead him down to the underworld may be seen, if you wish, as prefiguring his own death (the poem was written near the end of his life when he was ill with the tuberculosis that was to kill him), but the truly lovely invocation of 'blue darkness' saves it from any hint of the morose or the morbid. And whenever I see cornflowers now, they glow with that added layer of meaning for me, which Lawrence's poem has infused into my mind: they too might be torches to the underworld.

But if cornflowers are about darkness, the other blue flower I love, the harebell, is right at the opposite end of the spectrum: it is noted for its pallor. In fact, the paleness of harebells is part of their attraction, which is about delicacy. Occasionally confused with bluebells − they are of similar size, but whereas the bluebell is a hyacinth, related to irises and orchids, the harebell is a campanula or bell-flower, distantly related to the daisies − they are flowers of the end of summer rather than of the springtime, and while the bluebell's massed ranks are overpowering and unmissable, harebells can be overlooked. Sometimes you find them in small clumps; often they're just in ones and twos. Skimpy, skittish things, they are altogether frailer plants than bluebells; while the latter, growing in the rich damp soil of a woodland, have a fat sturdy stem which is bursting with sap, harebells, which flourish on dry open ground − I first got to know them on sand dunes − have a stalk which is just a wire. The bell-like azure flower on top of it could be made from tissue paper; it might have been cut out and pasted together by a child in primary school. This frailty means that it picks up the slightest puff of wind, quivering and nodding and catching the light in a continuous flicker. Christina Rossetti wrote:

Hope is like a harebell trembling from its birth . . .

The frailty and the flickering are among the points people notice immediately about the flower: a light show in the wind, a friend of mine once said.

(Christina Rossetti, by the way, was not the only nineteenth-

century woman poet to refer to harebells; there is also a harebell poem by the American Emily Dickinson, Rossetti's exact
contemporary – they were born within a week of each other
in December 1830 – which is so unusual and forceful, especially
in the shock of the minor but unmistakable erotic charge of
the opening, that I cannot resist quoting it:

> Did the Harebell loose her girdle
> To the lover Bee
> Would the Bee the Harebell hallow
> Much as formerly?
>
> Did the 'Paradise' – persuaded –
> Yield her moat of pearl,
> Would the Eden be an Eden,
> Or the Earl – an Earl?

It might take a bit of deciphering, but what Dickinson is saying
with her characteristic compression is that things which are
cherished because they are pursued, may be cherished no longer,
once attained.)

The deepest attraction of the harebell, however, is not so
much the flickering light show as it is the combination of its
colour with its timing. Its flimsy, pale sky-blue stands out because
when it appears, at the end of the summer, much of the life
has gone out of the landscape; the grasses have yellowed and
browned, the birdsong is silenced, the swifts have departed and
the trout no longer rise. There are flowers in bloom, such as
the pinkish-brown hemp agrimony, and harsh yellow ragwort,
but somehow they are part of a palette of exhaustion. The
calendar says what are you complaining about, it's still summer,
but I've always felt that summer really ends about 15 August,
and after that it feels like post-coital depression in the natural
world – a sort of in-between nothingness before the arrival of

autumn, with its own sharp identity. Into this time of melan-choly (for me, at least) pops *Campanula rotundifolia*: on heaths or dunes, grasslands or hillsides, the translucent blue bells catch the wind, catch the light, and catch the heart, with a colour which somehow seems to speak of the future rather than the past, even though everything around is starting to fade; they give the landscape a last flare of life at the point when the year begins to wither and die.

Both of these blues draw me powerfully, the pale defiant one and the dark pulsating one, yet neither can compare with the bluebells, for there is another blue which goes beyond and has a quite electrifying effect on me and I imagine on other people too, and which the bluebells contain, in their shade-shifting: it is the extreme, dazzling blue which nature very occasionally offers, a tone in which the basic colour intensifies itself more than any other colour seems able to, so that it becomes one of the most remarkable visual phenomena of the natural world. I think of it simply as brilliant blue. The examplar of it, for me anyway, is the blue of the morpho butterflies of South America, but it can be found in Britain in two other organisms besides the bluebells, both winged, one of which is a butterfly of our own, the Adonis blue, something of a morpho in miniature: it has the same lustrous brightness, the same glossy radiant sheen, on a much smaller scale. The seven blue butterfly species found in Britain are all pleasing, in fact, and one of the most attractive is the common blue, *Polyommatus icarus*, with wings of an irides-cent lilac, which I feel is probably under-appreciated because of its dismissive name – were it called the Icarus blue, say, people might cherish it more – but the Adonis blue edges it, in its brilliance. The first time I ever saw one, I was called over by a friend: the butterfly was resting on the turf, wings upright and closed, showing only the spotted brown undersides, and I crouched down and looked, heart beating, as he touched it with his fingertip, and there was a tiny explosion of blueness.

The other species is the kingfisher. The key point about kingfishers is that they bear two blues. One is the glowing greenish-blue of the folded wings, the colour seen in standard illustrations of the perching birds: it makes a striking and splendid contrast with the rich chestnut-orange of the underparts. But it is the other blue which takes your breath away. This is the blue of the kingfisher's back, and it is the blue you see in real life, rather than on the painted teapot or the greetings card, because your first sight of the bird is almost certain to be of it zooming away from you, and when its wings are outstretched, the back feathers are exposed and there it is.

It's a blue so bright it appears to be lit from within.

It's brighter than the sky.

This is not on any colour chart I have ever seen in a paint shop, and I feel that, as with the large copper, many people seeing this for the first time might have a sort of elated experience, which is that their sense of what the world can contain is actually expanded. That was certainly the case with my son Seb, when the two of us went for an evening walk while on holiday in Normandy, when he was seventeen. We were strolling along the side of the river which runs through the Perche, the Huisne, at an isolated spot near one of the many Percheron Renaissance manors, in this case the Manoir de la Vove; in the gathering dusk its fairy-tale towers stood out against the evening sky. The river ran between high banks and as the gloaming deepened, a blue light suddenly shot along the dark water beneath us and Seb pulled up sharply and exclaimed: 'What was *that*?' I told him; he was fascinated. It was a moment Seamus Heaney captured precisely with a line tossed off in a song of haunting music:

> I met a girl from Derrygarve
> And the name, a lost potent musk,
> Recalled the river's long swerve,
> A kingfisher's blue bolt at dusk . . .

Seb's is not a generation which looks at the natural world, but the blue bolt at dusk stopped him dead in his tracks.

As did the bluebells with me. In that wood, in that spring not long ago, for five days in succession I was struck dumb by the beauty of the earth. For five days I went back purposely to look at that colour, that living colour, because when I accidentally came across it, it was at its peak, and I knew that soon it would fade. Day after day after day after day after day. And I told no one. I think I was . . . what? Ashamed? No, not at all; but I am influenced by prevailing cultural norms as much as the next person, and I suppose I felt that declaiming about five successive days of bluebell-peeping would be regarded as eccentric? Or something? Yet I was drawn back there ineluctably, to glut my senses on colour. Without telling a soul. It felt almost like being part of the underground . . .

For if the beauty of nature is not high in official cultural favour, as we set out into the twenty-first century, it still holds its magnetism for countless unpolemical minds, with a force which strongly suggests it is rooted in our underlying bond with the natural world, and that culture is being trumped by instinct. That is certainly the case with me. I do not care a fig that modernism may have cast beauty aside, and that the legacy of the rejection may be with us today; to me, the beauty of the natural world retains its joy-giving power and its importance undiminished by artistic, cultural, or philosophical fashion – indeed, its importance is increased immeasurably by the fact that now it is mortally threatened.

And as for blue, and its special attraction for me, I think this is instinct trumping culture too; I know I am drawn to it beyond other colours, but I cannot see that I was socialised to be so in the course of my own life. If we accept that the human imagination was formed by our interactions with the natural world, over the fifty thousand generations, then I think that my blue-love is in there somewhere; that I probably have, planted within me, in the genes, a bond with a colour which was for our wandering ancestors the most predominant hue of all, stretching over their heads so far and wide that eventually they called it heaven.

The beauty of the earth, of course, goes far beyond colour. It is found just as much in form, both in its landscapes and in the life it hosts: in the harmony of vistas, the majesty of mountains, the intimate charm of valleys, and the changing light of the sun upon them all; in the killer grace of leopards, the elegance of antelopes, the dash and fire of falcons, or the poise, as I have said, of wading birds. I admire all of these, but there is one form, one type of landscape feature in particular whose beauty has given me joy, and that is rivers. Not just any rivers, though: a specific group of rivers, in a specific place, whose beauty is such that, to me, it almost seems to reach beyond the material world into the realm of the ideal.

I start from a point of prejudice in their favour: I have loved rivers all my life, or at least since I was entranced, at the age of eight, by the first threatened species I encountered, those gnomes of BB's Folly Brook. Their amiable watercourse with its unending murmurings and plops and splashes, its weirs and watermills as it swelled and widened, its hidden anchorages and overgrown islands, was very much the fifth main character of *The Little Grey Men*, especially when Dodder, Baldmoney, and Sneezewort, accompanied by their lost and now found brother, Cloudberry descend it towards the sea, desperately seeking a new life, in the book's sequel, *Down the Bright Stream*. Ever after, I have been unable to see a river, any river, without a quickening of the spirit – if I cross a river on any journey I want to know its name, and if at all possible, stop on the bridge and gaze into its currents – and yet, this is such an automatic reaction that as I have got older, I have wondered if perhaps this fascination too may be hard-wired, if this too may come from long before my childhood, from the hunter-gatherers; if perhaps the story of the Folly Brook may have switched on a pre-existing longing, one which was already there, deep in the tissues.

For although we may increasingly take them for granted the more urban and enclosed our lives become, rivers constitute one of the key elements of human existence. They did not have to be: just as we could have easily evolved on an earth without flowers, so we could have done on a planet without flowing water, and it takes an effort of the imagination to register what a singular phenomenon that actually is, eternally changing and eternally the same; we need Heracleitus to remind us, that no one steps into the same river twice. But being such notable entities, and being there in our evolution from the beginning, rivers in due course became for humans part of the very nature of things, and I remember the thrill of recognition I felt when I first saw this truth nobly expressed by Norman Maclean, the American professor of English and fly fisherman whose auto-biography became a celebrated Hollywood film. 'Eventually,' Maclean wrote, 'all things merge into one, and a river runs through it.'

So rivers are as much a part of where we come from as the sky is; but there is a discrimination to be made. They divide into two distinct categories: the giant rivers of the world, and all the others. The giant rivers seem to me to be a separate sort of creature from the rest, not only geographically, but also in our cultural responses to them, for they are much more than scaled-up streams; as water bodies on which voyagers may travel for thousands of miles, they are really lengthwise oceans. Go to the *encontro das aguas*, the meeting of the waters near Manaus in Brazil, where the Solimões and the Rio Negro unite to form the Amazon, their brown and black currents flowing separately side by side, and you will find that the Amazon has a horizon, just as an ocean does, and indeed, the giant rivers of the world defied exploration more than the oceans ever did: Europeans found the far side of the Atlantic long before they located the source of the Nile. It is the giant rivers, of course, which in early history most exercised the human imagination. Many of the first great civilisations coalesced around them: Egypt with the Nile, Mesopotamia with the Tigris and the Euphrates, India with the Indus and the Ganges, China with the Yellow River and the Yangtze. These stupendous watercourses were miraculously life-giving, but they could be furiously life-threatening; they could bestow riches but they could destroy in anger – the deadly Yellow River most of all – and the peoples who depended on them naturally made them deities to be worshipped, thanked, and placated; even in the twentieth century, T. S. Eliot, who grew up in St Louis on the banks of the Mississippi, could not help but see his colossal childhood waterway as 'a strong brown god'.

I have felt that awe of the Mississippi too, and of other giants such as the Niger, from the air at Timbuktu a great green ribbon of rice paddies snaking through Mali's yellow-brown semi-desert, at its surface the bearer of enormous painted canoes, the pirogues, symbols of its own enormous dimensions. But awe is not the same as love; and the rivers which I have loved have without exception been the others, the lesser ones, those on a human rather than a superhuman scale. How lame is English, for once, not to have separate names for them! French does it, naturally; a giant river is a *fleuve*, a lesser river a *rivière*, but in English they are pitifully lumped together, so the 2,900-mile-long Congo and the 85-mile-long Avon of Shakespeare are categorised by the same term. Let me be clear, then, that henceforth, in speaking of rivers, it is the Avons I mean, not the Congos; it is the Avons which have captured my heart.

The reason is, I think, that they are not there to be feared and placated, the smaller rivers, they are there to be befriended; and having my strong sense that all rivers are special anyway, like all butterflies are special, and that their individual differences are simply a magnificent bonus – as are their names – I have spent much time river-befriending, and always been rewarded, and I have fallen in love with many of them, from the Hodder, astonishing you that Lancashire, cradle of the Industrial Revolution, can contain such a jewel, to the Dysynni, dark and brooding and isolated under Cader Idris; I have loved bully-boy rivers such as the Helmsdale in Sutherland and sweet shy rivers such as the Lyd in Devon, and in particular, small rivers with literary associations, like Housman's Teme, or the Taw and the Torridge of Henry Williamson (and of Ted Hughes, too), or Dylan Thomas's Aeron (he and Caitlin named their daughter Aeronwy), or Seamus Heaney's Moyola, flowing down from the Sperrins into Lough Neagh, the river of the 'kingfisher's blue bolt' song quoted above, 'pleasuring beneath alder trees'.

They are all sources of delight, but the rivers which have given me joy, joy as I have tried to define it – the group of rivers whose beauty is such that it sometimes seems otherworldly . . . well, they are elsewhere. But not in some Shangri-la. You can find them on a map. Although it's a fairly uncommon one. You might have to order it. It is 'the 10-mile map' – the British Geological Survey's geological map of Britain, at the scale of 10 miles to the inch, which shows the country not in terms of administrative regions, or landscape features, but of the under-lying rocks. The various strata are variously coloured, for differ-entiation rather than resemblance (although the dull terracotta which marks the Triassic sandstone of the Wirral where I grew up is remarkably similar to the stone itself), and what excites me whenever I unfold it is the brilliant band of green splashed diagonally across England from the south-west, at bottom left, to the north-east, at top right.

It is the chalk. The green on the map represents the soft

white rock of the chalk hills, stretching from Dorset up through Wiltshire, Hampshire, and Berkshire, then into the Chilterns, and on to Norfolk, Lincolnshire, and the Yorkshire Wolds; it is formed from the remains of trillions of tiny marine organisms which filled the warm seas when the dinosaurs ruled on land, and whose shells settled on the seabed when they died; it is pure calcium carbonate, and is one of the great givers of beauty and wildlife richness. Often referred to as downs or downland, the chalk hills epitomise the tender charm of the southern English countryside. Their form is gentle and flowing (like the contours of the human body, it has been said) and quite unlike the paternalistic craggy dominance of the granite mountains of Wales and Scotland. Even more, they host incomparable bio-diversity, from the flower carpet of the short-cropped downland turf filled with scented wild thyme and horseshoe vetch and milkwort and fairy flax and orchids in abundance, to the butterfly parade of dark green fritillaries and marbled whites and silver-spotted skippers and blues galore, and to the birdlife from stone curlews to skylarks: on the chalk grassland of Salisbury Plain, for example, there are fourteen thousand pairs of skylarks, even today, and in spring they pour down a shower of song which seems as much a part of the air as the wind . . . But most of all, the chalk gives us its waters.

If I had to single out something to represent the beauty of the earth, one aspect alone, it would be the chalk streams of southern England. Their loveliness is dreamlike. They are small-to medium-sized rivers, but anglers long ago christened them the chalk streams, and such they have remained. Anglers have been their champions, their guardians, and their celebrants, fly-fishermen most of all, as they are regarded as the perfect trout streams, and in the literature of fly fishing, which is substantial, two of them occupy pride of place, the Test and the Itchen, both in Hampshire: on them Victorian anglers developed the technique of fishing the dry fly, which became almost a cult. A few more are celebrated in these writings: the Frome and the Piddle in Dorset, the Wylye and the Avon in Wiltshire (not Shakespeare's Avon but the one which flows through Salisbury), the Kennet and the Lambourn in Berkshire, the Chess and the Misbourne in the Chiltern Hills. Although there are many more smaller streams, including numerous rivulets you could leap across – the Environment Agency has counted 161 of them in total on the chalk belt running up the country – it is this handful of medium-sized rivers whose beauty sets them apart.

Their most striking aspect is the water itself: it is the cleanest and clearest river water on the planet, the customary epithet being 'gin-clear'. It is of a clarity to take you aback: the river bottom, which often in chalk streams is a lustrous golden gravel, is so perfectly visible it is as if you were seeing it through a single pane of polished glass. The reason is the geology. Because the chalk is permeable, it allows rain to soak through it to underground reservoirs or aquifers, and simultaneously filters it; when springs return the water to the river, all the impurities have been removed: the water is immaculate. This process allows for the chalk streams' second key characteristic, their constancy of flow. When rain runs off the land directly, as into so-called spate rivers, the water level can rapidly rise and fall; but because the chalk streams are spring-fed, their level is unvarying and their flow is stately, never sluggish, never torrential, with an elegance all of its own (the Test is like the Loire in miniature).

Thus, there is a beauty of essentials. This is further enhanced by the life the chalk streams are filled with, from the profusion of aquatic wild flowers above, led by the ranunculus or water crowfoot, the buttercup decorating the surface with its white stars and emerald leaves, to the fish underneath, such as that most vivacious and animated of fish, the brown trout. *Salmo trutta*. The salmon's junior partner. Exquisite, with no need to be gaudy like something from the tropical coral reefs: this is the restrained splendour of the north, boreal beauty, streamlined beyond the dreams of art deco. Ever watchful, ever on the alert. Eternally a-quiver, holding position in the flow ('On the fin', anglers say). And so marvellously visible in the gin-clear water, rising to take the upwing river flies on the surface, above all the mayfly, the largest and the prettiest, butterfly-sized and muslin-winged, living most of its life as a larva in the gravel of the riverbed and then hatching in late spring to mate and die in a single day. Thousands of them. The males perform a courtship dance, gathering in swarms and bouncing up and down, twelve to fifteen feet in the air; the approaching females are grabbed, and inseminated, and then lay their eggs in the river and expire in the surface film. And when that happens on any sort of scale, especially in the evening, the trout go mad: they attack the dying insects in rocket-like surges, rising to lacerate the surface in paroxysms of greed. Splash after splash. A watercourse pulsing with life and death.

I discovered the chalk streams thirty years ago from walking alongside the River Chess in the Chilterns, and as I began to explore them and to realise what they were, I was astonished, first at their phenomenal, head-turning beauty, and then at how minimally they seemed to be appreciated, outside the culture of angling. In fly fishing and its literature, these rivers were given their due, but beyond that, they might have been on the moon. Poets did not sing them. Painters did not paint them. Writers did not write about them, even nature writers who wrote about many aspects of the countryside, unless they happened to be fishermen too, like Viscount Grey of Falloden – he who remarked that the lights were going out all over Europe – who wrote as lyrically about the Itchen as he did about birdsong. The chalk streams seemed, and still seem, to have no place in the national consciousness. To me, they stand without question alongside the bluebell woods – these are the two supremely beautiful features of the natural world in Britain – but I have no sense that people widely share this view, or that they share my feeling that rivers like the Test and the Itchen are great national monuments which should be cherished as much as our medieval cathedrals.

It seemed remarkable, that they were so disregarded; but I didn't mind. I knew what I was looking at. I felt as if I were in on a secret that only anglers knew about, and I began to devour the literature (books like Harry Plunket Greene's *Where the Bright Waters Meet* or John Waller Hills' *A Summer on the Test*) until at one stage I became a near obsessive, travelling to see them all. For example, I have followed the courses of all the principal tributaries of the Test – the Bourne Rivulet, the Dever, the Anton, the Wallop Brook and the Dun – driving down lanes, peering over bridge parapets, walking by water wherever I could. And gradually, as I got to know them, I began to understand what was truly exceptional about the chalk streams – even, to employ the much abused word, what was unique. It wasn't just about beauty, it was more than that. It was about purity.

I would contend that the archetype of pollution, as a modern phenomenon, is the polluted river. Not that the pollution of standing water bodies, or of the oceans, or of the land, is one whit less to be concerned about; but somehow I feel we have in our minds, when pollution is spoken of, a primary image of flowing waters that have been dirtied or defiled – an image we do not like. Large-scale pollution is fairly new, in historical terms, and is a very much more recent environmental difficulty for the earth than deforestation, say, dating back less than two hundred and fifty years to the Industrial Revolution. In that initial explosion of no-holds-barred capitalism, rivers were the natural world's first victims, being brought into bondage by the first factories, to provide power and to take away waste. They continued to be sullied and despoiled until the widespread collapse of manufacturing industry in the West in the 1980s, since when, not a few have been cleaned; but throughout the nineteenth and the major part of the twentieth centuries most factories and industrial complexes, and most industrial towns and cities in the western world, had an associated river, which was filthy. Many millions of people will have seen such a watercourse. (Now large-scale manufacturing has shifted eastwards, and it is China which is taking river pollution to new heights: alas, poor baiji.)

Yet I do not think people are indifferent to this, even if it does not affect them directly; I think we instinctively find polluted rivers very unpleasant, as a concept as much as an empirical experience. I suggested earlier that we may have an attraction to rivers which lies deep in the genes, a part of the bond of the fifty thousand generations, and if so, then the attraction is clearly to rivers which flowed long before large-scale industrial pollution arrived to blight the earth; to rivers which were pure and could themselves be purifiers, since they could take away human wastes and not be polluted by them, when our numbers were small: these rivers were truly worthy to be cherished. If, then, there is an image of a river buried somewhere in our tissues, and I think there may well be in mine, it is clearly of a wholly untainted one, almost of a Platonic ideal, one that we long for; and thus to see a river despoiled causes distress, even if we can't quite say why.

But what, in the modern world, could ever match the internal image of purity? What, in mere material existence, can approach the ideal? We might travel the globe over and never find it, and that is very probably what happens with most of us; unless, that is, we chance upon the chalk streams. They come as a shock. For suddenly, the ideal is real; the internal image is matched. It is hard to convey the faultless nature of their water. Indeed, in *A Summer on the Test*, John Waller Hills, early twentieth-century Conservative politician and fisherman, wrote that one day the water of the Anton, the Test tributary, seemed to him 'unimaginably pure'. It is not only purer than you have seen before, or purer than you would see anywhere else, it is purer than you would allow yourself to expect or even conceive of, and so it begins almost to seem not part of everyday life, but of some ultimate condition; and the purity of the water casts its glow over the river as a whole, as an entity, and the river too seems to be something other than an everyday river, it seems almost to belong to a higher state of existence.

Hyperbole? You could say so, I suppose. But what can I do, other than speak of my experience? Once, on a May morning a few years ago, I came out on to the banks of the Upper Itchen, at Ovington in Hampshire, and the river with its flowers and willows and the serenity of its flow and its dimpling trout in its matchless, limpid water, all gilded by the sunshine, seemed to possess a loveliness which was not part of this world at all.

Yet it was part of it; and there, once again, was the joy.

There is one more river I wish to instance, in looking at the joy we may find in the beauty of the earth, but this case is different, for it's about failure: it's about where joy might have been but is not. It's about a dream which in the end could not be realised; yet I feel the story is worth telling, on many levels, not least as the river is one of the most famous in the world.

It is the Thames: London's river, and my own, or at least, that's how I think of it, as I have lived near it for more than twenty years, am fascinated by its history, and ride my bike every week for miles along its towpath, watching its changing waters and its changing moods. It is a handsome as well as a historic river, not least in the part I know best, the eleven-mile stretch from Hampton Court, past Teddington and Richmond to Kew, which runs through a green valley on London's edge dotted with at least nine great houses or stately homes: this is the nearest thing we have in Britain to the châteaux of the Loire.

Biologically, however, it is no chalk stream. The Thames has suffered from some of the grossest river pollution Britain has ever seen, and two hundred years ago this led to an extinction which bears direct comparison with the current case of the baiji in the Yangtze, that of the Thames salmon. You may not think of the Thames as a very salmony river, yet until the beginning of the nineteenth century it harboured a significant population of *Salmo salar*, the Atlantic salmon, of the great ocean-wandering fish which came back to breed; and although there appears to be no historical basis to the often repeated story that London apprentices got so tired of being fed salmon they had it written into their indentures that they would be served it no more than once a week, there is no doubt that salmon catches from the Thames were substantial, with as many as three thousand fish a year being taken to Billingsgate market, and individual netsmen frequently making sizable hauls: on 7 June 1749, for example, 47 fish were taken in a single day below Richmond Bridge. There were big fish, too, with several records over 50 lbs, while 16 lbs was a good average. It was a solid, self-sustaining salmon run, dating back thousands of years.

Yet in the blink of an eye – in historical terms – it was wiped out. Compared to other cautionary tales of extinction, such as that of the dodo or the great auk, the story of the disappearance of the Thames salmon is unknown to the general public, but it is as egregious an example as any of the terminal effects of mankind's activities on living things. It was extremely rapid, taking little more than twenty-five years, and pollution was its cause.

For centuries, London's river had received waste material but had been powerful enough to flush away whatever the inhabitants of the growing city threw into it, and thus it remained more or less ecologically sound. However, a point eventually came when it was overwhelmed. After 1800, as the Industrial Revolution took off, London's population began a mammoth surge, rising from 960,000 in 1801 (the first national census) to 1.6 million in 1831 and 2.3 million in 1851. Two aspects of this boom were fatal for the salmon of the Thames. The first was the vast increase in the amount of sewage going directly into the river, especially after cesspit overflows and house drains were allowed to be connected to the public sewers (which until then had essentially been drainage ditches) in 1815. London's human waste, its 'nightsoil', which for centuries had been collected by cart and spread on the land as manure, began to pour into the river just as the population exploded, and the process was given a further savage impetus by the contemporary invention of the water closet.

The second was the burgeoning industrialisation of the capital and the consequent discharge of toxic effluent into the river from the factories mushrooming along or near its banks, not least from the new gasworks which were built in increasing numbers after gas lighting of London's streets began in 1807. The waste discharge from gasworks was of exceptional toxicity, containing a cocktail of noxious substances ranging from carbolic acid to cyanide. With raw sewage slushing in on the one hand and lethal contaminants on the other, the Thames in London started to turn into a great poisonous, stinking ditch.

But there was a third contemporary development which told against *Salmo salar tamesiensis*, as we might call our beast: this was the building on the river, upstream of London, of pound locks and their associated weirs, to facilitate navigation of heavier and heavier loads to and from the industrialising city. It happened quite quickly. Teddington Lock, which immediately constituted a new limit for the tidal Thames (historically, it had been Staines) was built in 1811, Sunbury in 1812, Chertsey in 1813, and Hampton Court in 1815. These locks and their weirs became formidable barriers to migratory fish trying to get upstream; they also changed the whole nature of the river as the water built up, deepened, and slowed behind them, and the natural gravel shallows in which salmon could spawn silted up and disappeared, or were dredged away.

The salmon were doomed. They could not live in the filthy water; it was much harder to swim up the river out of it; and even if they could, they could not breed. Their headlong demise is graphically illustrated in the melancholy record of salmon taken at Boulter's Lock, Maidenhead, by the Lovegrove fishing family between 1794 and 1821. In 1801, 66 fish were caught; in 1812, 18; in 1816, 14; in 1817, 5; in 1818, 4; in 1820, 0; in 1822, 2; and that was it. George IV particularly sought a Thames salmon for his coronation feast on 19 July 1821; none could be procured. The last Thames salmon is believed to have been caught in June 1833, cited (although with no location given) in William Yarrell's *A History of British Fishes*, published in 1836.

There were no more for a hundred and forty years. After the fish disappeared the pollution of the river continued to worsen, until it finally became insupportable: in July 1858 a heatwave made the smell from the Thames so hideous that Parliament at Westminster was compelled to stop sitting. This famous episode, known as 'The Great Stink', led directly to the construction of the modern London sewer system by the engineer Sir Joseph Bazalgette. Bazalgette shifted the problem away from central London by building giant interceptor tunnels on either side of the river – the Thames embankments were built to house them – which carried the capital's sewage eastwards, about a dozen miles downstream of Tower Bridge, to outfalls at Beckton, on the Essex side of the river, and Crossness, in Kent.

But although this relieved Westminster and the City of the worst of it, it merely transferred the pollution somewhere else; the outfalls were not sited far enough downstream for the ebb tide to flush the untreated sewage right out to sea, and it returned on the flood. This massive 'plug' of nauseating filth in the lower river proved mortal to any fish in its vicinity. Throughout the rest of the nineteenth century and the first half of the twentieth the situation continued, and after the war years if anything worsened, until in 1957 a survey carried out by the foremost authority on British fish, Alwyne Wheeler of the Natural History Museum, proved an astonishing point about the fish of the tidal Thames: there weren't any.

Between Kew in the west and Gravesend in the east, Wheeler established, there were no viable fish populations. This survey severely jolted the public conscience, and after a couple more damning scientific reports on the chemical and biological state of the river, what should have been done years before was finally carried out: from 1964 the Beckton and Crossness sewage works were cleaned up and the effluent from their outfalls treated before discharge, so that its micro-organisms were no longer capable of sucking all the oxygen out of the water, which had been its most damaging characteristic. The effect was virtually immediate. From the mid sixties onwards, fish started to re-appear in the river. They were monitored by Alwyne Wheeler, who had the bright idea of asking the power stations along the banks to check on what was being trapped in their cooling water intake screens. Starting with a tadpole fish (a member of the cod family), species after species began to turn up – lampern, sand goby, roach, barbel, John Dory – until by 1974 no fewer than 72 species had been recorded. And then the incredible happened: on 12 November 1974 an 8 lb 12 oz salmon, a hen fish 31 inches in length and four years old, was caught in the intake screens of West Thurrock power station, near Dartford, about sixteen miles below Tower Bridge. Wheeler himself examined it and identified it that day; and that day, the dream was born.

I often cycle up the towpath on the Surrey side of the river from Richmond to Teddington Lock, and cross over, to return down the Middlesex side; and as I push my bike over the footbridge I gaze at the weir, the weir that marks the limit of the tidal Thames, and I see in my mind's eye the flashing silver fish, shouldering its way upstream, driven by the imperious urge to reproduce, driven up and over that foaming barrier, leaping for life . . . what a creature the salmon is! What would you not give to have it back in your river! Would that not offer you joy? Many people thought so, when Alwyne Wheeler pronounced his identification: the gates of possibility seemed to swing open. There was huge excitement, enormous publicity – here was the first Thames salmon for 141 years! – and people's thoughts almost immediately turned to what might be. Was the river really now so clean that the ancient salmon run might be resurrected? Encouragement was given by two more finds of salmon penetrating further and further upstream: the remains of a 21-inch-long fish found on the foreshore at Dagenham in July 1975, and then – quite remarkably – a fish found dead where the River Mole joins the Thames, at Thames Ditton, on 30 December 1976.

It was remarkable because it was above the tidal limit, that is, above Teddington Lock.

It must have jumped the weir . . .

So the Port of London Authority set up an inquiry to look at migratory fish in the Thames, and when this eventually reported that conditions in the estuary would no longer be an obstacle to returning salmon and sea trout, and there was now a good chance of reintroducing both, and all the stakeholders were duly and fully consulted, and all the relevant committees had met, and all the 'i's were dotted and the 't's crossed, in 1979 the Thames Salmon Rehabilitation Scheme was established and the dream became official. I have always thought it a tremendous dream. Rarely can a piece of public policy have enshrined such an exhilarating vision: this legendary, righteous fish, which, with its need for a high dissolved oxygen content in all the fresh water it passes through, is such an icon of aquatic purity, was to be returned to the river as the supreme symbol of the Thames reborn, of a great watercourse brought back to life. This is the fish which speaks to us of Scotland, of Norway, of Iceland and Nova Scotia, of northern wild places which are wholly unspoiled, and London's river was to host it again. London's river was to be a salmon river: could ambition be any nobler?

That this did not happen, even after thirty years of trying and the expenditure of a colossal amount of thinking, of devoted effort and of cash, is one of the saddest things I have witnessed in all the struggles to shore up the natural world against the depredations we have made upon it. Yet the first part of the recovery programme was a big success. Its aim was to prove that fish could pass down to the sea through the once impossibly polluted estuary and successfully return back through it. They did, in eye-catching numbers. Smolts – juvenile fish six inches long – stocked in the tideway (that is, below Teddington Lock) were eminently able to come back after a winter or two winters at sea, being captured in the salmon trap specially installed at Molesey weir near Hampton Court. In 1982, 128 returning fish were trapped; in 1986, 176; in 1988, 323; and in 1993, the peak year, 338. This was the period of newspaper headlines about the returning Thames salmon, which reached a climax with the first rod-caught fish, a six-pounder taken by Mr Russell Doig from Staines, in Chertsey Weir pool on 23 August 1983. Mr Doig won the silver cup and £250 prize offered by the Thames Water Authority for the capture, which had earlier been claimed by at least two other anglers, but not awarded (draw your own conclusions . . .). He was pictured in a TWA photo with his rod and his fish (the fish looking stiff from the freezer) in a boat in front of Tower Bridge. It was a staged but unmistakably telling image, which shouted London and shouted salmon, and as such was priceless publicity; yet in so far as it reflected the clean-up of the tideway, the message it was giving out was a true one.

However, the tideway was only half the story. The salmon were returning to spawn. But where? There were no suitable sites left in the main river. After an intense decade of trials, the best potential spawning ground was identified in the upper reaches of a chalk stream tributary of the Thames in Berkshire, the Kennet; attention focused on the gravelly bottoms of an isolated stretch of the Kennet known as the Wilderness Water, where stocking with fry – baby salmon – was begun. Yet from the Wilderness to Tower Bridge was seventy-five miles of river, and this second long passage, above London, proved much harder for the fish to negotiate, each way, than had the seventy-five miles of once polluted estuary, between London and the sea. As soon as stocking began on the spawning grounds, rather than in the tidal river, the number of returning salmon dropped dramatically. Not the least of their problems was that between Teddington Lock and the Wilderness Water were no fewer than thirty-seven weirs. Even though some fish might jump some weirs, from the point of view of the salmon's supporters, choosing the Wilderness meant that every one of those weirs would have to have a fish pass installed. And so, over a period of fifteen years starting in 1986, this was done at a cost of several million pounds, the money mostly being found by the Thames Salmon Trust, the charity set up specifically to raise it. This seems to me an astonishing and very largely unsung achievement, which climaxed with the opening of the last fish pass to be built on the Kennet, on the weir at Greenham Mill, Newbury, in October 2001.

Even though the experimental restocking of Thames salmon had been continuing for more than twenty years, it was not until this moment that a proper self-sustaining salmon run once again became a real possibility. And yet it has not emerged. No fish, as far as is known, have to this day – thirteen years on, at the time of writing – swum all the way down to the sea after being stocked as fry in the Wilderness Water, and then swum all the way back up and bred in their 'home' stream, which was the whole point of the exercise – bar one. A single salmon is known at least to have made the journey each way. It has a name, or, rather, a number. It was found in the salmon trap at Sunbury weir on 14 July 2003 by Darryl Clifton-Dey, the Environment Agency scientist who was running the Thames salmon scheme, a cock fish weighing twelve-and-a-half pounds; it was estimated to have spent two years at sea, and was recognised as a 'millennium baby' – one of ten thousand tiny fry stocked in the Wilderness on 9 June 2000. Darryl and his colleagues put a small radio tag on it bearing the number 00476, and released it; and the following November, on the 28th, they picked up its signal in the pool below the weir at Hamstead Marshall – weir number 37, the gateway to the Wilderness Water – thus proving, to their great delight, that the arduous spawning journey for Thames salmon is indeed possible. Even if it hasn't happened.

But 00476 and the proof of his odyssey – had he gone to Greenland? – were not enough. For another eight seasons the Environment Agency persevered with stocking salmon fry in the Wilderness, but no breeding was observed and only the odd fish returned to the lower Thames each summer; and at length, after the stocking of 2011, the Thames Salmon Rehabilitation Scheme was brought to a close. It had lasted thirty-two years.

Having followed it fairly closely for more than two decades, I have thought a great deal about its failure, about the reasons for it, and about the lessons to be learned.

There is no doubt that over the years of the scheme, circumstances moved against the fish and fresh problems arose, two in particular. One was that the increasing amount of abstraction from the river by water companies to supply their customers in hotter years was making the flow often insufficient to tempt salmon upstream from the estuary, something which may worsen further with climate change; the other, more immediately serious, was that in London itself, Joseph Bazalgette's ageing sewer system was increasingly discharging raw sewage into the river when heavy rainstorms filled the pipes completely, which led to rapid deoxygenation and mass fish kills. This latter phenomenon has become known as 'London's dirty secret', and in September 2014 the British government authorised a solution for it, the building of a new £4 billion 'supersewer' to intercept all the rainfall discharges, which is due to be completed in 2023. Will that help salmon back to the Thames? Perhaps.

The lessons, for me, are about our limits. We grow used to wildlife conservation success stories, of endangered species miraculously brought back: in Britain we have the sea eagle, the lady's slipper orchid, the large blue butterfly . . . further afield we see the American bison, the Arabian oryx, or what about the Mauritius kestrel, there were only *four* of those left at one stage and now there are hundreds of them . . . we know we are wrecking nature across the globe, but those of us concerned with conservation have generally tended to feel that if conservationists direct their efforts at saving a specific species, with a fair wind and enough funding, they can usually succeed. Well, not always. The principal lesson of the Thames salmon story, for me, is that we can sometimes damage the natural world too severely for it to be repaired.

Yet more than the lessons, more than the reasons for the failure, what I take away from it is the sadness. It was a dream, perhaps with a hint of the Romantic about it, but an eminently practical one – assuredly it was an inspiring one – and to watch it die is a heavy weight on the heart; although, maybe it is just the project that has died, and not the dream. Certainly, crossing the footbridge over Teddington Lock, pushing my bike, I still see him in my mind's eye, the silver shadow below bulling his way upstream towards the tumbling water of the weir; and then I also realise something more about the beauty of the earth, as he leaps, that it is found not just in colour, not just in form, but in life itself.

7

Wonder

Since the beginning of this book I have barely referred to butterflies, but sixty years ago they did indeed fly into my soul; and they have never flown out. For a long time, I did not know how to categorise what happened to me in Sunny Bank, at the age of seven, since experiences which mark you for life, as a child, generally arrive as dire or at least, disturbing ones, be they physical or psychological; yet this experience, although it came in a time of turmoil, was not in itself distressing, although it was most assuredly powerful – and that I have always understood, as it has exercised a form of dominion over me ever since. It was as if it had implanted something permanent into my nervous system: a receptiveness to butterflies which was almost a whole new instinct. A lepi-empathy, if you like. It has been a peculiar part of me, a quirk of personality like a limp or a lisp, like a quick temper or a meanness with money, and it will last until I die. I say again, this did not mean I became a butterfly obsessive – I was not Fowles' Frederick Clegg and there were long periods when I gave no thought to Lepidoptera, turning my adolescent enthusiasm to birds, as I have described – but it did mean that there has always been present in me the possibility of an intense response to butterflies, especially when the encounters were unexpected.

All down the years.

For example, it happened in April 1968, when I was twenty and a student at the University of Toulouse, and was spending the Easter holidays hitching around Italy to look on the monuments of the Renaissance, and I had been to Florence and seen what everybody sees, but most particularly the young Lorenzo de' Medici, who was one of my heroes, portrayed on horseback in the Gozzoli fresco of the Journey of the Magi to Bethlehem – there was still a tide mark across the painting from the calamitous flood of the Arno eighteen months before – and I had marvelled at the Piero della Francesca *Resurrection* in the little civic museum in Sansepolcro, and I had admired Federigo da Montefeltro's palace looming over Urbino, and a French couple in an Alfa Romeo – funny the details you recall – had driven me down from the hills and then on to Rimini, where I slept on the beach. I had run out of money. With the last of it I had bought two loaves and half a dozen eggs, which I had hard-boiled in the youth hostel in Arezzo; and the next morning, which was a Saturday, I remember, I made my way to the Rimini entrance to the autostrada and took stock. I had three eggs left and a loaf, no cash, and about seven hundred and fifty miles to hitch, back to Toulouse, but I reckoned I could do it easily enough, over the Alps from Turin down to Marseille where the widow of a friend of my father's lived, and she might bale me out, and anyway it had been worth it, God yes . . . I had stayed in the castle in Lerici from where Shelley set out and was drowned, and I had squeezed into Savonarola's cell, and I had discovered the portraits of Bronzino which excited me more than anything else and then I saw the swallowtail.

It was on a roundabout, a traffic island – one which had not been landscaped with lawn or flower borders but was simply in-filled with autostrada construction rubble and thus was full

of weeds or wild flowers, however you want to describe them
– and it was catching the bright spring sunlight of the Marches.
I forgot about the Renaissance. I forgot about hitch-hiking
logistics. I was electrified. Here was something from the far
corner of my imagination, one of those early dream-species
from *The Observer's Book of Butterflies* which I had continued
willy-nilly to dream about and which in Britain, then, was the
rarest butterfly, as it was confined to one small area of the
country, the Norfolk Broads. It was also the biggest butterfly
species, but even more than that, it was . . . I find it hard to
get away from the word *glamorous*. It was the most glamorous
British butterfly, and I had never set eyes upon it until this
moment in north-eastern Italy.

By glamour I suppose I mean something like beauty with
built-in excitement. Clichés beckon. Movie stars. But there was
undoubtedly something in this insect's appearance, in its banana-
yellow slingback wings slickly transected by bold jet-black stripes,
which set it apart. Its was not a calming colour scheme. It was
flashy, it carried a hint of risk, even of danger, and today, with
innocence long gone, I might say there is even something almost
tarty about it, as if the pair of black needletails to the hindwings
were stiletto heels. I was mesmerised entirely. I watched it intently
for perhaps three or four minutes, until it finally flew off and
I waved it goodbye, excitement slowly subsiding, and planted
my Union Jack-decorated rucksack at my feet and got out my
thumb. That night, as I climbed into my sleeping bag under a
pine tree on the outskirts of a small town near Alessandria called
Tortona, about seventy miles short of Turin, my left lung collapsed
and the events which followed changed my life, but when I
think back to that day, the swallowtail is what I remember first.

I could tell you something similar from nearly a decade later,
from May 1977 when I was in Rondônia in the Brazilian
Amazon as a reporter for the *Daily Mirror* – the pre-Robert
Maxwell *Daily Mirror* which had tried to become the *Guardian*

for ordinary working people and which I loved and believed in – and for a *Mirror* series called 'The Last Frontiers' I was writing about the settlers swarming over the rainforest like ants on the carcass of an elephant, in the first great wave of Amazonian deforestation. They were grim young men from the south in straw hats, everywhere slashing and burning the trees so that the landscape of flaming and smoking stumps looked like the aftermath of an armoured battle. I was focusing on the fact that they were encroaching on the territories of Indian tribes, some of which had only just been discovered or contacted, and which FUNAI, Brazil's national Indian foundation, was trying with difficulty to protect. I was engrossed by it all, this rising human tide, this vast irresistible surge of destruction which a decade later was to obsess the world but which then was only just beginning, and at the same time I was engrossed by the thoughts of a woman thousands of miles away in America, a woman who was forty-three while I was twenty-nine, a woman who, although I had been in love before, had presented me with my first experience of passion, and everywhere I went I saw her heart-stopping face and her fire-red hair, in Rio, then in São Paolo, then in Brasilia, and then in Porto Velho as we reached the Amazon, and then in the tiny frontier towns and settlements as we got deeper into the jungle until eventually we found ourselves, four of us – myself, the photographer, the interpreter, and the guide – at a log cabin at the end of the farthest path from the farthest track you could possibly go down. The cabin had been put up several months earlier by a settler and it was in the territory of the Suruis, an Indian tribe which had only been contacted three years before, in 1974, and it was illegal, as it was built in the Surui reservation which FUNAI had demarcated, the demarcation post being clearly visible back down the narrow path through the rainforest along which we had hiked after leaving the Land Rover. We talked to him, the settler – his children had a Surui arrow, we noticed – and he was staying, he wasn't going anywhere, and a rifle was hung behind the door, and he cheerfully took us across a fallen-tree bridge over a small river to show us the fruits of his labours, a patch of virgin jungle half the size of a football pitch which he had cleared himself and planted with bananas, and I realised that this was the sharp frontal point, at that moment, of the human invasion of the Amazon. (And even there, I saw her face.) Beyond, beyond the packed impenetrable trees, was the Surui village, ten miles away, maybe fifteen, nobody really knew because the only way you could reach it was to fly to the small airstrip FUNAI had built, but they weren't going to take us so how could we get there – that was the problem. This was a great story but it was only half a story; we had to go and see the Suruis themselves – that was what I was pondering as we left the banana clearing and scrambled back over the fallen-tree bridge and said goodbye to the settler in his cabin and headed back down the track to the vehicle as quickly as we could because the rain was coming and that would make the track impassable, how could we get to the village, that was the question, and the morpho flew out of the forest.

I stopped dead in my tracks. I had never seen anything like it in my life. Never such a creature. Never such a living thing. A substantial chunk of the pure blue cloudless sky which had fallen to earth. It was probably, I suppose now, *Morpho peleides*, or maybe *Morpho menelaus* – I can't make a judgement because I was too stunned to note down any details. I was rooted to the spot, even as the others turned and shouted, *Mike, come on, the rain's coming!* for it was enormous, not only in physical size but in its blueness, its iridescent metallic blue, its brilliant blue, its burning blue, its incandescent blue . . . I forgot about the deforestation of the Amazon. I forgot about the Suruis and their inaccessible village. I even, beshrew me, forgot about the heart-stopping face, for the first time in weeks, and nearly four decades later, when I close my eyes and think of the rainforest and the morpho flutters out of the trees, I smile when I realise that, even if momentarily, it actually eclipsed the passion, and I would have said to you, then, I would have sworn to you, that nothing, nothing whatsoever, could do that.

Over the years I have had a number of such encounters, such as my first silver-washed fritillary, floating down through the oak woods of the Haddeo valley on Exmoor, and my first Camberwell beauty on a forest track in Provence, and my first monarch in a Boston garden – they might be common to Americans, but they aren't to me – and what has characterised them all has been intense emotion, a feeling almost of being struck dumb, and I have gradually come to understand it, and to realise that this is what I experienced in Sunny Bank; and its name is wonder.

In a book about joy, this is a digression of sorts, but a necessary one, for wonder is the other great feeling which nature can trigger in us, and that we might experience it, seems to me even more remarkable than the fact of the joy experience. I have written about it, fairly briefly, in the past: I once described as a sense of wonder the emotion felt by myself and my then eleven-year-old son in listening together to a nightingale sing a few feet away from us, deep in a wood at midnight; and I think that many people may have experienced such a feeling in their encounters with the natural world, and been greatly moved, without perhaps understanding exactly what the sentiment is. To explore it therefore seems worthwhile, as my instinct is that wonder as much as joy may show us the way to our nature bond, which was forged in the psyches of our distant ancestors and is surviving in ours today.

Today, though, wonder is much like joy in the popular mind,

in that it is discounted: in a secular and sceptical age, it is not a notion we have much to do with, and it figures little in our everyday discourse. Yet it is there in the repertoire of human feelings as much as it ever was. It is related to joy, perhaps, but there are significant differences: one, that it is harder to define. There is no doubt that joy is a concentrated happiness, however we characterise its overtones, while wonder is a concept on which thousands of words have been expended, without a generally agreed definition being arrived at.

The *Concise Oxford* has a useful stab at it: 'An emotion excited by what is unexpected, unfamiliar, or inexplicable, esp. surprise mingled with admiration or curiosity'. I would put it rather differently: I would say wonder is a sort of astonished cherishing or veneration, if you like, often involving an element of mystery, or at least, of missing knowledge, but not dependent upon it; for true wonder remains when the mystery is no more, or when the missing knowledge is supplied. It involves 'an astonishment which does not cease when the novelty wears off' (Kant, quoted by the British philosopher Ronald Hepburn).

My sense of it is of an emotion by which we are overcome, comparable to the religious experience, on the one hand, or the aesthetic experience, on the other, and it signifies that there is something very special to us about its object, perhaps through what that object makes us feel about our place in the world. I think deep down the feeling is, that we are astonished to be in a world which can contain such a phenomenon – the nightingale singing in the darkness, say – and somehow, the astonishment then reaches out beyond the sense of our place in the world, merely, to the fact that we exist at all. Human existence is taken for granted virtually all the time, of course, it is the greatest of our complacencies, but experiences of wonder can jolt us into the realisation of how remarkable not only our own but all existence actually is – *Why anything? Why not nothing?* – and an arresting illustration of this was given by Ralph Waldo

Emerson at the start of his essay, *Nature*, with a flight of fancy as charming as it is vivid: 'If the stars should appear one night in a thousand years, how would men believe and adore; and preserve for many generations the remembrance of the city of God which had been shown!'

And yet we do not have to glimpse the full glory of the universe to experience wonder: it can be triggered in our personal lives by art (by classical tragedy especially), and by spiritual epiphanies (rare as hen's teeth now), and more to the purpose here, by aspects of the natural world. Let me give another example. In June 2004 I took my two children, Flora and Seb, on a half-term holiday to a fairly remote Greek island, Alonissos in the Sporades (my wife Jo had to stay behind at the last minute because her father had fallen dangerously ill). Flora was twelve and Seb was just coming up for his eighth birthday, and one morning the three of us joined a trip on an old-fashioned local boat, a kaïki, in the past used for transport and fishing but now pressed into the service of tourism. We were bound for the even remoter island of Kyra Panagia, to visit its ancient monastery and have lunch, and in the sunshine the silvery-blue Aegean was quite unusually still: by the time we were halfway there the surface was as glassy and flat as I have ever seen it, without a ripple or a wavelet, it was a true mirror-calm over which the kaïki dreamily glided in the heat haze, and the dozen of us who were passengers were relaxed by it to the point of sleepiness, when the water next to the boat exploded. A pod of common dolphins, half a dozen of them, suddenly surged out of the sea to look at us and play around us. Every soul on board cried out in amazement and delight, and as we looked on spellbound, they performed an acrobatic display in and out of the water for three or four minutes, all about the boat. Then they simply disappeared, and the sea was flat as a millpond once again.

The whole company was stunned. It was hard to take in exactly what had happened. We seemed to have been subject to a visitation: these large, strikingly beautiful, fierily energetic creatures had purposely come to see us, out of nowhere, and they seemed to have intelligence, and friendliness, and even an exhilarating sense of fun, and we realised at that moment what increasing numbers of people have realised over the last thirty years or so, from the decks of boats: how singular are the cet– aceans – the whales and dolphins – and in particular, how extra– ordinary they can seem, in their interactions with us. They may be wonder-inducing, above all other animals; and in exploring wonder in nature, they are a good place to start.

The appreciation of their unusual qualities is very recent in the rich industrialised West and represents a fascinating cultural shift, but one not often remarked upon as such, since it is hard to categorise. Under what rubric do we discuss it? Psychology? Zoology? Tourism studies? Whales and dolphins figure strongly in ancient folklore, of course, especially in the legends of peoples who lived near the sea. In Genesis, they were the first animals to be brought forth by the Lord – 'and God created great whales' – while for the Greeks, dolphins were among the stars of the natural world, fresco favourites, major mosaic motifs, and the stories of them saving people from the waves figured not only in myths (and on coins) but also in serious history; Herodotus recounts, as something to be believed, the story of Arion the poet, who was tossed overboard by the treacherous sailors of his Corinthian ship and brought to shore by a friendly dolphin. But in the farming and then industrialised culture of modern Europe and America, down the centuries, cetaceans played virtu– ally no part, other than in Herman Melville's strange and spell– binding *Moby-Dick* (published in 1851 but not widely read before the 1920s), until a series of events in the post-war years brought them out of obscurity and into a new folklore of our own times.

The first was the vogue for performing dolphin shows which swept the rich world from the early 1960s, inspired by the

Hollywood film *Flipper* and its spin-off TV series; at one time there were no fewer than thirty-six dolphinaria, aquariums with dolphins, in Britain alone (by 1993 they were all gone from the UK, but according to a 2014 report, more than 2,000 dolphins, 227 beluga whales, 52 killer whales, 17 false killer whales, and 37 porpoises were still being held in 343 captive facilities in sixty-three countries – I can't imagine any of them are happy). The second was the emergence, in the 1970s, of new environmental pressure groups such as Greenpeace and Friends of the Earth, who began campaigning for an end to the commercial hunting of the great whales, with its egregious cruelty, crying out the archetypal slogan of the modern Green movement: *Save the whale!* As a result of these events, cetaceans swam into the modern consciousness, and there they have stayed, growing steadily more prominent because of a third development, the emergence from the 1980s onwards of organised whale watching.

The recreational observation of whales, dolphins, and porpoises in their natural habitat is now a substantial activity across the world: the most recent survey estimated that 13 million people took a whale-watching trip in 2008, and my family and I were among them. For after the visitation on the voyage to Kyra Panagia, we all wanted to see as much of cetaceans as we could, so whenever possible, whale watching was included in holidays. Over the years we managed to get close to Dall's porpoises and gray whales off the coast of Vancouver Island, to humpback whales off Cape Cod (with the spectacular sight of a humpback breaching, leaping clear of the water), to the bottlenose dolphins of Cardigan Bay (with a mother and calf coming close to the boat), and to a pair of spirited and vivacious common dolphins which had taken up residence in a sea loch in the Scottish Highlands; we have also enjoyed briefer glimpses of harbour porpoises and minke whales.

We were thrilled by them all. From the moment we caught sight of them, we were energised and excited, we thought they were special, even though we did not have the remarkable experiences of close contact in the water which people can sometimes be blessed with, and which make them feel that whales and dolphins are different in nature from all other non-human animals, and of a higher order, with jaw-dropping characteristics: not only their complete mastery of another world than ours, but their real wish to interact with us, their seeming intelligence, their playfulness, their friendliness and gentleness, their apparent, occasional singling-out for attention of people who are in some way troubled. To read about this in detail, or to talk about it at length, as I have done with Mark Carwardine, the naturalist and TV presenter who knows more about cetaceans than anyone else in Britain, is to enter a sort of no-man's-land between slowly developing science and rapidly accumulating anecdote. Formal research is making it increasingly clear that cetaceans are indeed exceptional in many ways: to take but two examples of many, dolphins are now known to possess among their vocalisations 'signature whistles', in effect personal names, raising the question of whether they also have self-awareness; while some bowhead whales are now believed to live as long as two hundred years, or even more. But it is the encounters with people over the last thirty years, encounters subject to no experimental protocols but often mere haphazard events, which are generating the real wonder and a new folklore. As a scientist who leads whale-watching trips, Mark is in the middle of it; he follows current research closely and is fully aware of the dangers of anthropomorphising, yet after many years of close observation he is in no doubt, for instance, that dolphins ride the bow-waves of boats, not, as some scientists would still maintain, merely as a way of getting from A to B, but simply for fun.

He wrote the world's best-selling guide to cetaceans; he is the man who presented the celebrated radio series (and subsequent book) about vanishing wildlife, *Last Chance To See* with Douglas Adams, author of *The Hitchhiker's Guide to the Galaxy*

(and later reprised it as a television series with Stephen Fry).
'I've run hundreds of trips, to see all sorts of animals all over
the world, gorillas, elephants, rhinos, tigers, and they all have a
big impact on people,' he said to me. 'But what I've noticed
over the years is that whales and dolphins have a different and
a greater impact. It's not just because I'm biased. I've seen it
time and time again.' One of the biggest impacts is when he
takes people to the San Jacinto lagoon in Baja California in
Mexico, where gray whales come down from the Arctic to give
birth to their calves, and where in the past they were slaughtered
by whalers. Now the females and their calves come up to whale-
watching boats to be stroked; and considering the history of
the place, the trust they display sometimes leaves the strokers
overwhelmed. 'People can be completely changed by the experience,' Mark said.

We did not enjoy this sort of intimacy, myself, my wife Jo,
and our children Flora and Seb, and yet the whales and dolphins
we did see on our trips undoubtedly filled us with a sense of
wonder. When I was first trying to analyse why, I was at something of a loss, but a conversation with Flora, the most enthusiastic whale watcher of us all (who was by now twenty-two),
opened doors in my mind. She said: 'They're like beings from
a different dimension.' I was much taken by this, and we talked
about it further, and in the end I asked her to write her thoughts
down so I could remember them.

She wrote:

Why I like whales is something to do with the fact that
they are 'other-worldly' – manifested in their physical
strangeness (i.e. so big, so slow, out of time with the rest
of nature) – and almost a throwback to the dinosaurs. Their
other-worldliness relativises and undermines our world view
– i.e. life is richer/stranger than we remember on a daily
basis. There are other hidden dimensions (e.g. the ocean)

that are just as much a part of the earth, but which are so forgotten by us on a daily basis, and quite literally, invisible to us, as the deep ocean has no sunlight.

She concluded: 'So, whales are so magical because when they surface, they offer a physical/visible token of another realm which is veiled from us, but which also comprises part of our planet.'

In other words, they offer mystery; there, is one of wonder's prime sources.

There are a number of triggers for wonder, in the natural world; for example, besides mystery, we may readily observe a couple of specific conditions, diametrically opposed but equally capable of sparking amazement infused with delight, and they are rarity and abundance; yet there are also some aspects of nature which are less obvious, but which, when encountered, can produce wonder at the very existence of the earth and our existence upon it. One of them is simply the age of things: so very much has gone before us that it cannot be justly computed; rather, all that can be registered is the scale of it:

> Very old are the woods;
> And the buds that break
> Out of the brier's boughs,
> When March winds wake,
> So old with their beauty are—
> Oh, no man knows
> Through what wild centuries
> Roves back the rose.

Walter de la Mare understood it: the extent of All That's Past. And there is another less familiar facet of nature which can also seem wondrous: its ability to transform. The idea of transformation is one of the most resonant of our imaginings: we are fascinated by people changing identities, by things becoming different things, by frogs which turn into princes. Shakespeare lives on such stories; Ovid's *Metamorphoses* was not only a bestseller in the Rome of Augustus, it was probably the most popular book both of the Middle Ages and the Renaissance. Obviously, there are different directions which transformations can take, including the tragic, the humorous, and the ironical; but it seems to me that the two basic ones are down and up. Down is the transformation of ill-fortune, of being changed from banker to beggar, of being King Lear losing everything; but surely the transformation which most appeals to us is the transformation upwards, when people or creatures or things which are merely mundane become special, or even resplendent. That idea seems to strike a deep chord within us, to touch some primal longing. It is much more than the idea of gaining wealth or status, or even the idea of the ordinary girl who becomes a princess, say; it is something at the heart of myths and religions, including Christianity, the notion that with all our faults, we might aspire, silly though we know the idea is, to perfection. And one spring I gave a lot of thought to this, in trying to understand the effect on me of a particular phenomenon of the natural world: the dawn chorus.

For several weeks I had been trying to finish a long piece of writing, and to do that I had taken to working through the night. If you work through the night you see the dawn. Or rather, you hear it. At eight minutes past four in the morning of 21 May that year, a sound came to my ears; I stopped typing, got up, and went and opened the kitchen door to the back garden. Light was flooding the eastern sky, a great rising tide of pale light, although the surrounding houses and trees were black silhouettes against it; a misty moon still shone; there was no wind, only an absolute stillness; and from the top of a tall copper beech tree two gardens away, liquid and clear on the air, a blackbird was singing.

There was no other sound. The blackbird sang his unending phrases as if the stillness were intended specially for him, for they were floating on the quiet, every one precise, hypnotic in their music and their purity; and then, from a nearby rooftop television aerial, a second blackbird joined in. Shortly after that there was a robin; then a blue tit; then a goldfinch: the dawn chorus had begun.

I do not know – no one is quite certain – exactly why songbirds all sing together at first light and then fall silent (they are likely to be either proclaiming their territories or inviting mates); but I do know that, having gone out to listen to it a dozen times in the weeks which followed, as it got earlier and earlier (until one morning it began at 03.34), it is entrancing. At first I thought it was simply the symphony of birdsong itself which moved me so much, but now I know it is something else as well: its transformative power. For I live in suburbia. I live in a land of neat gardens, estate agents' boards, car ports, walked dogs, lawnmowers, endlessly similar houses and nothing much happening, a land which no one would ever describe as resplendent. Yet the dawn chorus clothes suburbia in wonder. Like the visits of Father Christmas or Roald Dahl's Big Friendly Giant, it takes place while most of us are still asleep and so we miss it, and I felt, after those few weeks, as if I had discovered a secret: that in the chorale of birdsong silvering the silence, the stillness and the great bursting dawn overhead, for a brief half-hour even the land of the lawnmower can approach perfection.

But finding wonder in the mundane is exceptional: transformation is indeed required. Finding it in the mysterious, as in the whales and dolphins which come to us from a different dimension, as it were, is much more likely, even though the most notable characteristic of mystery today is how fast it is shrinking. We regret this. Mystery matters to us. Which is curious, for, if we define it as the phenomenon of not knowing, it has in the past been a principal cause of stress, as not knowing is something to which humans temperamentally are not indifferent; it is probably at the basis of religion. The otter does not worry that its river may dry up in a drought, as far as we know. But we worry. What causes that disease? What causes those crops to fail? Will my future be good or bad? Who are we, and why are we here? Once possessed of consciousness, human beings were never able to sit calmly alongside such ignorance, but had to do something about it; and what can you do, other than dream up all-knowing, all-powerful supernatural beings, whom then you may propitiate?

Since the scientific revolution of the seventeenth century, however, we have been steadily eroding mystery, and now we generally know what causes that disease or causes those crops to wither, if not, alas, what form our future will take. Yet mystery's retreat from our lives is by no means widely welcomed, for, paradoxically, we seem to be as much attracted by it now as we have been frightened by it in the past, and we regret its disappearance (and you could write a very saleable tome on the Decline of Mystery, with a cover featuring, say, Neil Armstrong's bulbous boot treading the moon's mystery down). Now that its terrors are lessened, mystery appears fascinating; it seems to have definite qualities, not least, to appeal to the problem-solving part of our nature, which I suspect goes back to the fifty thousand generations, just as much as the fear of the unknown does. 'Mystery has energy,' says the character Conchis in John Fowles' *The Magus*. 'It pours energy into whoever seeks an answer to it. If you disclose the solution to the mystery you are simply depriving the other seekers of an important source of energy.' It also confers an undoubted allure. Wouldn't you like to be thought of as mysterious? I know I would (some hope). And so with the natural world.

We are hugely drawn to mysteries in nature. Does the supposedly extinct ivory-billed woodpecker still exist in the wild woods of Arkansas, as the Cornell Laboratory of Ornithology so loudly proclaimed in June 2005, to headlines around the world, or does it not? Since no one has been able to repeat the Cornell Lab's sightings, a decade on, and the greatest American expert in bird identification says the blurred video put forward as conclusive evidence shows a pileated woodpecker, not an ivory bill, the question mark over the claim is very large; but that only enhances the mystery. We are gripped by it. I am gripped by it, anyway. I was completely gripped by the talk I had with one of the Cornell team, Melanie Driscoll, who said to me: 'I've been treated like a fantasist, and I've been treated like a rock star, but I know what I saw.' I have been gripped by such things ever since, as a young man, I first came across a book by a Belgian zoologist which reverberated through my imagination, and does so still to this day.

First published in 1958, it was called *On the Track of Unknown Animals* and its author, Bernard Heuvelmans, examined the idea that there were still large, wild creatures left to be discovered, and that some of them might be remarkable relics from the past. Heuvelmans' book formalised an area of enquiry – an enthusiasm, if you like – which came to be called cryptozoology: the search for animals whose existence has not been proven. Cryptozoology has unfortunately morphed into pseudo-science, fixated as it is with the Loch Ness Monster, the Yeti, and the Bigfoot of the American north-west (now generically referred to as 'cryptids'), not to mention ABCs or alien big cats – seen the Surrey puma lately? Or the Beast of Bodmin? – and indeed, it quickly shades into preoccupation with UFOs and the paranormal, and features prominently in publications such as *Fortean Times*, specialising in weird news.

So far, so wacky. But Heuvelmans himself was a classically trained zoologist (his doctoral thesis was on the teeth of the aardvark) and his book is a scrupulously sober amassing of information not only on beasts which are formally unknown to science, but also on animals which had been unknown but had been discovered fairly recently, such as the pygmy chimpanzee (the bonobo) and the Komodo dragon, the giant monitor lizard of Indonesia, as well as on creatures which had gone extinct in the recent past, such as the thylacine, or Tasmanian tiger. The method is far from fantastical; it is straightforward assessment of evidence in numerous cases, showing that this one had been found, and that one had disappeared, and this other one was perhaps waiting to be discovered. Some of the cases he highlighted were potentially sensational and have greatly excited adventurers; other suggestions were more restrained, and some of them seemed to me plausible.

One such was that of the woolly mammoth, which we generally think of as dying out tens of thousands of years ago, but which we now know (from carbon dating of its remains) survived on Wrangel Island, off the coast of Siberia, until at least 1650 BC. Heuvelmans' proposition, based on scattered, obscure reports from Russian hunters, was that isolated, relict populations of mammoths might still survive in the taiga, the endless conifer and birch forests of the Siberian mainland. We are far more familiar with the Amazon rainforest, yet the Siberian taiga is bigger, with enormous areas wholly unpenetrated by roads, even today; and I thought when I read his suggestion, and I continue to think: why not?

For in our arrogance we assume we have mastered the natural world; yet its power to surprise us persists. Even though we have substantially shrunk the unknown part of the planet, and mystery with it, I rejoice in the fact that in my lifetime there is still enough of it left to contain creatures of which we have no knowledge, and which are wondrous in their discovery. This is the case with two habitats especially: the remaining rainforests and the deep oceans. In recent years the rainforests of Indochina in particular have been bounteous in yielding up big unknown animals, largely, of course, because war put the jungles off limits for so long to explorers and naturalists. The most spectacular has been the ill-fated Vietnamese rhinoceros, discovered in 1988, poached to extinction by 2010 – nobody knew beforehand there were any rhinos in mainland Indochina at all – but we have also been vouchsafed the saola, or Vu Quang ox, a sort of cross between an antelope and a buffalo with long, backward-sloping horns and a white-striped, mournful-looking face, discovered in 1992, as well as at least three new species of Vietnamese deer. Meanwhile, just since the start of the new millennium, alongside numerous fish and other organisms, two completely new species of whale have swum into our ken (Perrin's beaked whale and Deraniyagala's beaked whale), while one previously known only from skeletal remains, the spade-toothed whale, has been seen in the wild for the first time; and there will surely be more.

We may well feel wonder at them. We certainly would were the woolly mammoth to reappear, or the ivory bill. In fact, I am not sure which, in the end, would seem the more wondrous, since although the mammoth might have a prehistoric air, when all is said and done it is more or less just a shaggy version of the Asian elephant, while *Campephilus principalis* is so dazzling a creature that it was known as the Lord God Bird, as those lucky enough to catch sight of this acme of all woodpeckers sometimes could not refrain from crying out *Lord God!*

But even creatures still definitely present in the world about us have mystery which may shade into wonder, and one has long possessed my imaginings: the Clifden nonpareil. This is a moth whose name proclaims that it is without equal entirely; and I agree. It is the most magnificent moth to be found in the British Isles. Not only is it enormous, in moth terms, it has another characteristic that sets it quite apart from our other

867 larger moth species: blueness. That hue again! Moths generally favour browns and greys, though sometimes red and yellow and orange and cream, and occasionally green; but apart from the odd fleck, such as in the eyespots of the eyed hawk-moth, blue as a colour is, from the British moth fauna, missing completely.

Yet not from *Catocala fraxini*. Its other English name is the blue underwing, and when it opens its silver-grey, delicately patterned forewings it discloses underwings (or hindwings) of black, each crossed by broad bands of a stunning smoky lilac. It's a shock, and that's the point. There are a dozen or so 'under-wing' moths in Britain and all employ the same use of colour as defence-through-startling, in a way I referred to earlier, with the Jersey tiger. The forewings are cryptic, perfectly camouflaged to blend in with a daytime resting place, a stone wall or the bark of a tree, but if a predator such as a bird does notice it, the moth will suddenly snap its forewings open and display an underwing colour flash, bright enough to confuse the bird for that extra second to help it get away.

Some of these species, such as the large yellow underwing, are common; but not the Clifden nonpareil. No sir. It is not only exquisite, it is very rare, with only a few sightings each year, and it has been hugely prized by all those interested in Lepidoptera since it was first observed in the eighteenth century at the Thameside estate of Cliveden in Buckinghamshire (where Nancy Astor famously held court between the wars, and where in 1961 the Conservative minister John Profumo met the young would-be model Christine Keeler and began the sexual scandal which forever bears his name). It's been a legendary treasure, a holy grail for moth enthusiasts. Mystery hovers about it.

For myself, a Lepidoptera lover and an unashamedly nerdy moth man, I dreamed about it, I longed to see it for years and years, without success. I thought I never would. But one autumn, in early October, the charity Butterfly Conservation (BC) announced that there had been an influx of rare moths from the Continent, and this included a spate of Clifden nonpareil sightings, three of them by one of BC's moth experts, Les Hill, in Dorset. A day or two later I was Dorset-bound, to see if Les could produce one for me. When I arrived at dusk he had his moth trap out in his garden and I was prepared for a long vigil, but Les had startling news; his colleague Mark Parsons, the charity's head of moth conservation, had half an hour earlier actually found a Clifden nonpareil on the wall of his cottage, thirty miles away, and caught it. He had it still. We drove pell-mell to the other side of the county and eventually, in a plastic box in Mark's kitchen, there it was. Asleep. (Do moths go to sleep? Well, torpid, then.) Motionless but miraculous. For when Mark gently touched the silver-grey forewings, they shot open, and there were the bands of that glorious lilac-blue. I couldn't believe I was seeing it.

It began to stir then, and shortly began to fly slowly around the kitchen. I was open-mouthed. It was as big as a bat – a bat of sensational colours. Eventually it settled on the kitchen wall, and before Mark released it, I persuaded it to crawl on to my hand. It felt like a dream. The hyperbole cannot be helped. Astonishment at the world, that it can contain such a thing.

I have looked at *how* we may feel wonder in nature through my experiences, as you may doubtless feel it sometimes through your own. I could offer more, from rarity to abundance, from gazing spellbound on the lady's slipper orchid, the single plant that for fifty years and more was the rarest organism in Britain, looked after by dedicated carers in complete secrecy, to marvelling at the profusion of life still remaining in the countryside of Romania, as yet undefiled by intensive farming: the hay meadows overflowing with wild flowers (twenty-seven species in the first one I looked at), the hillside grasslands more spectacular still, above Viscri in Transylvania, where I walked through millions of blooms of dropwort and yellow rattle forming an endless carpet of white and gold which was crowded with insects like shoals of fish in the sea (grasshoppers and crickets, exquisite beetles like the rose chafer, stunning butterflies like the poplar admiral, the Hungarian glider, and the clouded Apollo), and the bird fauna just as rich, with red-backed shrikes everywhere and golden orioles giving their fluting whistles from the poplar trees (and there were bears in the woods . . .). But *why* we may feel wonder, is a yet more fascinating question.

To me, the ability to experience it implies some sort of preexisting relationship between us and the natural world; that is, an inherent one. The perception of the object of wonder does not fall on infertile ground. There is some faculty inside us which receives it and engages eagerly with it, and Wordsworth realised that with his 'sense sublime / Of something far more deeply interfused',

> Whose dwelling is the light of setting suns,
> And the round ocean, and the living air,
> And the blue sky, *and in the mind of man* . . .

Something dwells already in our minds; and I believe it is the bond, the bond of fifty thousand generations with the natural world, which can make aspects of nature affect us so powerfully: as with joy, so with wonder. The wonder in Sunny Bank is a case in point; when I looked up at the buddleia bush, I had not been socialised to react to the beings I saw upon it. You may say, you must have seen books about butterflies at school, or pictures at least, and perhaps I did; but if I did, I have no memory of them. I was seven years old. I did not feel, Ah, these are the butterflies about which I have read so much. I merely reacted to what was in front of me. And I have noticed, throughout my life, that virtually all small children react strongly to new creatures. They become absorbed in them instantly. They are very rarely indifferent. It is another human universal. That was the reason why zoos were so successful, until they fell out of favour – *Daddy's taking us to the zoo tomorrow, we can stay all day*. It was not books, or pictures previously seen, which bound me to the butterflies in Bebington. But it may well have been that observer in prehistory, my distant hunter-gatherer ancestor, who waited for a swallowtail to settle, the better to look upon it, and then marvelled at what was there in front of him.

After I came back from South Korea, from having borne witness to the destruction of Saemangeum, in April 2014, I wanted to know how the spoon-billed sandpipers which had so depended on the lost estuary were faring, the ones which had been brought to Britain in the attempt to establish a conservation breeding programme, and so I got in touch with the Wildfowl and Wetlands Trust in Gloucestershire, which was looking after the birds. Its director of conservation, Dr Debbie Pain, invited me to Slimbridge to see them.

It was an enormous privilege. The biosecurity around the specially built aviary was formidable, and Debbie herself could not accompany me, as she had a cold, a potentially lethal danger to the 'spoonies'; so after various scrubbing routines and the donning of sterile overalls and clogs, I went inside with Nigel Jarrett, the head of conservation breeding at WWT and the man who had godfathered all the birds since they were eggs in remote Chukotka.

This was another occasion for wonder, to be suddenly in among this flock of some of the rarest birds on the planet:

preciousness personified. There were twenty-five of them: tiny, bright-eyed and graceful, and quite unafraid, they were incessantly active, foraging around my feet through the aviary's imitation tidal pools; they were just coming into breeding plumage, exchanging the grey of winter for the russet head of their lovely summer dress. Nigel pointed out the increasing skittishness of their behaviour, chasing each other, one male raising a wing as a territorial warning, and all diving for cover when an oyster-catcher – perfectly harmless – flew overhead calling. Previously they'd been living together as a flock and had been quite happy with each other's presence, he explained, 'but now there's a surge of hormones going through their little bodies, causing them to moult, and be interested in each other. They're like highly strung teenagers.'

When we left the aviary we went back to Debbie's office to discuss the breeding programme and we talked widely of wildlife and the natural world. Debbie was an inveterate wildlife traveller and had just come back from a trip with her husband to Ladakh, where they had managed to see snow leopards in the wild, and Debbie said, a friend of hers had remarked that such a sight had to be the best wildlife experience ever, and she had been about to agree when she thought for a second, and said to her friend, *No. No, it wasn't.*

I was intrigued. 'So what was?'

Debbie said: 'Bioluminescent dolphins.'

'What on earth are they?'

Debbie explained that she and her husband had gone on a whale-watching trip to Baja California, taking in the San Jacinto lagoon – the very area which Mark Carwardine focused on – and she echoed Mark in her feelings for the cetaceans they saw, the dolphins especially. 'I love cetaceans,' she said. 'I feel a real connection with them somehow. Dolphins are amazingly joyful animals, they're forever leaping out of the water and going in the bow of the boat, and I'm a scientist, I think there has to be a reason for it, but actually I think they're doing it because they're just having a really good time . . . they're having . . . fun, you know? That's what I really want to think.'

One night, she said, the bioluminescence appeared in the sea – the green light given off, under certain conditions, by billions of tiny plankton (the older word for it is *phosphorescence*). It was a remarkable spectacle. 'The beauty of bioluminescence . . . it's such an incredible thing, when you get no moon and certain sea conditions . . . it's glowing green, it's stunningly beautiful to see, even if it's just coming off the spray. And to see shoals of fish in it, that's amazing. You can see these outlines and these streaks and these huge glowing green areas where a whole shoal comes towards the boat and then they dart off in all sorts of directions, leaving this glowing trail behind them – they look radioactive. It's just astonishing.'

But not as astonishing as the dolphins.

'We were in the boat, on the bow, and we saw these streaks coming in from the distance, coming towards us, we saw the outline of them, glowing green, and then they came up, below us, four feet below us, glowing there – and it just takes your breath away.' There were tears in her eyes. She said: 'I'm almost in tears being transported back to it. They played around the boat. Glowing green. They were just heart-stopping. It's the most amazing thing I've ever seen in my life. And the same for my husband. He just came up and grabbed my hand and said, we are never going to forget this, and we stood on deck for about an hour after they'd gone. Not wanting to go in. Not wanting to go to bed.'

She said: 'It was about midnight, when they came.'

Then she said: 'You know what the feeling was? Above all?'

'What?'

'What an amazing place the world is.'

She looked into the distance, lost in the memory. She shook her head, at the wonder of it. Then she turned to me and smiled.

'I think it will be the last thing I will ever remember.'

8

A New Kind of Love

What does it all add up to then, the snowdrops and the mad March hares, the coming of the blossom and the coming of the cuckoos, the bluebell woods and the chalk streams, the corn-flowers and the harebells, the magnolia warbler, the Clifden nonpareil, the salmon leaping the weir, and the mournful fluting of the wading birds drifting over the marshes – one person's lifetime delight in nature? I stand by what I said at the outset: defence through joy. But is that not in fact a hopeless idea, like some hippie in 1969 saying, we must win over the skinheads? How could an increasingly stricken planet, subject to the fiercest forces of destruction, actually be protected by human happiness? The answer lies in what the joy we find in nature may tell us about ourselves, and what that then may lead to.

Looking back over it all, I would say once more that the fact we might love the natural world, that we might give our hearts to it instead of just taking what it has to offer and tiptoeing around its pitfalls, still seems to me almost too extraordinary for words, and to mark us out as unique animals almost as much as language does, or the possession of consciousness. The sin-gularity of the basic premise, that we might love nature at all, never seems to be remarked upon: *why* can our feelings be engaged by the dawn chorus, or by dolphins visiting us from their different dimension? Compared with why people vote the way they do, say, or why people's attitudes alter with age, or why people sometimes commit murder, there is no investigation, the question is never put: does anyone ever go out with a clipboard and ask it? Yet these feelings are real, and remarkable.

It should be recognised at once, of course, that plenty of people, perhaps a majority, do not share them. But it is not in any way my contention that the love of nature is universal. What is universal, I believe, is the *propensity* to love it; the fact that loving it is possible for people. That propensity seems to me to be not an occasional trait to be found in certain individuals, but rather a part of being human, and a very powerful one: it is part of the legacy of the fifty thousand generations of the Pleistocene, our undying bond with the natural world, and it is no surprise that it lies buried in the genes, since it is covered over by the five hundred generations of civilisation we have lived through since farming began and we ceased to be part of nature ourselves as hunter-gatherers; and it is covered more than ever now by the frenzy of modern urban living. Yet covered only; not destroyed. It's still there. It can be uncovered; we can connect to it, all of us, and if we do, one realisation, one truth, may be illuminated for us more than any other: that the natural world is our natural home, it is the natural resting place for our psyches. And the most striking evidence of all for that is simple: it can bring us peace.

Having experienced a childhood in which there was disturbance, I am much taken with the idea of peace. I am only too aware that I belong to the luckiest of all generations, we who grew up in the rich West during a time of unheard-of, luxurious peace, after our parents and grandparents both had to endure world wars; but I suppose that the peace which particularly engages me is not peace between nations, crucial though that is, but peace at the minor scale: peace that may come, sometimes, to the troubled mind. For I believe it is possible; and I find haunting, and often think about, the refrain of the wonderful motet by Vivaldi, *Nulla in mundo pax sincera* – there is no real peace to be had in this world – as I disagree with it. (The idea, of course, is that true peace can only be found in Jesus, and it is the setting of an anonymous Latin poem which is actually quite bitter – it speaks of the venomous serpent found among the blossoms, and you wonder about what had happened in the life of the unknown author. Although it is ostensibly written to the glory of God, what actually comes through is the melancholy, which somehow pervades and adds to Vivaldi's exquisite melody.)

To me, peace *can* be found in this world, and in nature especially, as much as anywhere; and since Roger Ulrich, thirty years ago, published his arresting discovery that the patients recovering from surgery who could see trees through the hospital window, recovered far faster and more fully than those who could only see a brick wall, as I instanced earlier, we have begun to study formally the beneficial effects of the natural world on the human body and the human mind, and there is now a substantial literature. To illustrate nature's restorative powers, I could quote from it any number of examples, but following Joseph Conrad's belief that the influence of the artist is more enduring and goes deeper than the work of the scientist, I prefer to quote a poem. It's called 'The Recovery':

> From the dark mood's control
> I free this man; there's light still in the West.
> The most virtuous, chaste, melodious soul
> Never was better blest.
>
> Here medicine for the mind
> Lies in a gilded shade; this feather stirs
> And my faith lives; the touch of this tree's rind,—
> And temperate sense recurs.

No longer the loud pursuit
Of self-made clamours dulls the ear; here dwell
Twilight societies, twig, fungus, root,
Soundless, and speaking well.

Beneath the accustomed dome
Of this chance-planted, many-centuried tree
The snake-marked earthy multitudes are come
To breathe their hour like me.

The leaf comes curling down,
Another and another, gleam on gleam;
Above, celestial leafage glistens on,
Borne by time's blue stream.

The meadow-stream will serve
For my refreshment; that high glory yields
Imaginings that slay; the safe paths curve
Through unexalted fields

Like these, where now no more
My early angels walk and call and fly,
But the mouse stays his nibbling, to explore
My eye with his bright eye.

I have quoted it in full because I like it very much, and it is not well known, and I think it deserves as wide an audience as possible. It is by Edmund Blunden, one of the poets of the First World War, and what he was recovering from, was what he had seen in the trenches. Blunden went to France as a nineteen-year-old subaltern in 1916 and remained at the front for most of the rest of the war, far longer, in terms of continuous service, than all the better-known poets such as Wilfred Owen, Robert Graves, and Siegfried Sassoon; his survival was a miracle. But if his body escaped unscathed, the wounds to his psyche were monstrous. You can get a vivid sense of them from his account, *Undertones of War*; it is the most restrained of all the famous Great War memoirs, but even so the true ghastliness of what is happening in front of him, of bodies blown to pieces on a daily basis, cannot, willy-nilly, be disguised, and for the rest of his life in academia and the literary world he lived with the horrors, suffering regular nightmares. (His daughter Margi told an interviewer in 2014 that while Blunden's day was filled with being a literary journalist or a professor, 'the night was filled with the war.') What adds to the power of the poem is that the horrific memories are referred to only fleetingly and indirectly – 'imaginings that slay' – they rumble in the background, they are distant thunder, and the concentration is instead on everything that eases them, everything that is commonplace and reassuringly familiar about nature; above all, there is a sense of coming home.

Home indeed. For the natural world is where we evolved; where we became what we are, where we learned to feel and to react. It is where the human imagination formed and took flight, where it found its metaphors and its similes, among trees and pure rivers and wild creatures and grasslands rippled by the wind, and also among poisonous snakes and lethal predators and enemies and the unending quest for sustenance – but not among concrete buildings and automobiles and sewers and central heating and supermarkets, for these last are just accretions, add-ons which have been with us merely for the blink of an eye in evolutionary time, no matter how much they now may dominate our lives. Deep down, they mean nothing. It is nature which is the true haven for our psyches, and I was given a more detailed insight into why this is so by Nial Moores of Birds Korea, during the time I spent with him looking at the lost estuary of Saemangeum and the surrounding coastline.

It was fascinating to pass several days in the company of a

man who had spent long continuous periods observing nature, and especially shorebirds, with a very sharp eye. For this had given Nial a particular interest, which was in how wild creatures moved through a landscape and how they interacted with it; and his interest had broadened, from watching birds in a land-scape, to watching people. He had no doubt that we still possess the reactions to landscape of our Pleistocene forebears, such as the primal need of all wild things to see and not be seen; to see prey and avoid being seen by predators. He was fascinated by the way people react to open space, for example: they instinct-ively tend to walk around the edge, he said, where they are less conspicuous, rather than walk across the centre, where they are very visible.

Nial felt that as humans we were hard-wired to expect certain things from a landscape, such as a harmony, certain symmetries, an expected relationship between objects: 'a crown of a hill should be followed by the deep of a valley, which should be followed by another crown of a hill'. (Modern artificial land-scapes often violate this principle.) Thus, we were also hard-wired to process signals from it, in sights or sounds or smells, above all those which denote sudden difference or change. 'Something different is danger, something outside the expected harmony – it's the bear, it's the wolf, it's the stranger from a new valley.' This constant processing took a lot of mental energy, but we had evolved to do it over the thousands of generations; we were adapted to it. 'There may be danger, but it's a danger your body can understand.' But what we could not do, he said, was cope in the same way with the relentless stream of signals emanating from the city; we became numbed to the plethora of noises, lights, smells all around us. Because we were exposed to them constantly, it simply took too much mental energy to process everything that might be a threat, so we shut them all out, and that was the cause of stress. In the natural world, however, we could function once again as we have evolved to do.

I agree fully, and I would also add something of my own: that much as we may love it, the natural world is not paradise. Anyone who equates nature and paradise is missing the point. Nature can harm you and nature can kill you; nature indeed has dangers. But they are *our* dangers, as it were, and whatever they may be, they are part of the ecosystems to which, at the profoundest level of our personalities, we are all adapted; the earth's biosphere is simply what, as emerging humans, we grew unshakeably used to, over fifty thousand repetitions of 'the camping trip that lasts a lifetime', which is why it is still our home, and why it can bring peace to all of us – to her, to him, to you, and indeed to me. For I too have had my own coming to peace through the natural world.

It was an uncommon one I suppose, for it took all my life to arrive; but when at last it came, it set the seal on something I had longed to commemorate, as it was a meaning-making, for a very specific set of circumstances, which only nature could have provided. It concerned my mother Norah and her trouble, her three breakdowns, which took place when I was seven, nine, and eleven, and which devastated our family, but of which the single most remarkable feature – I now realise – was that Norah recovered completely.

I think this is unusual. I have nothing but sympathy for those who have suffered psychic shocks and are left with lasting wounds, in fact I suspect this may well be the norm: there is a disturbed state which persists for the rest of life. It was not so with Norah. Once she had finally regained her equilibrium, in the autumn of 1958, although naturally she bore the surface scars on a daily basis, being hesitant and insecure, it was almost as if – for a key part of her – the breakdowns had never occurred. The part I am referring to is what I would call her essence, her informed intelligence; there, there were no aftershocks, no resentment at what had befallen her, no seeking of sympathy, no jealousy of others, not a smidgeon of self-pity – rather, her essence remained unaltered, an essence which was, as I said at the outset, entirely unselfish, wholly honest, and gentle and kind to a fault.

Over the years of my adolescence which followed, I left behind the strange indifference with which, in stark contrast to John my brother, I had greeted her disappearances, and gradually began to build bridges with her. It was her intelligence which drew me in; I perceived it at first only dimly, but I found that as my mind expanded, and in particular when I fell in love with poetry at the age of fourteen, she was a matchless sounding board, explaining and reinforcing and opening my eyes to people I might otherwise not have encountered (it was she who introduced me to Gerard Manley Hopkins, for example). She was no less concerned, of course, with the development of John, who had come out of the years of family trauma very obviously damaged, and who for his distressed and anxious personality I still considered an embarrassment (which must have caused Norah much pain), and who had been further divided from me by that most inequitable feature of the 1944 Education Act, the Eleven-plus examination: while I 'passed' and went to a grammar school and studied languages and the humanities, he 'failed' and went to a secondary modern school and studied vocational subjects such as woodwork, metalwork, and domestic science.

But in his teens John began playing the piano (ours was a family with music running through it) and it was soon clear he had a notable talent, and when he took his Grade Five examination and won a distinction, the examiner said, if he can get some GCEs, this boy can go to music college – the General Certificate of Education then being the basic school-leaving academic qualification, but one for which pupils in John's secondary modern, unfortunately, were not entered. My mother went to see his headmaster and begged him to let John take his GCEs, to no avail: he left school at fifteen with nothing and Gordon, Mary's husband, got him a job as an office boy in his firm of ship's chandlers in Liverpool. But Norah would not accept defeat; withdrawing John from his office boy job, she bought a series of correspondence courses, through which, at home, over a period of two and a half years, she single-handedly taught him from scratch the syllabuses of five GCE subjects: English Language, English Literature, Music, French, and Religious Knowledge. In his examinations, he passed them all, and was accepted by the Royal Manchester College of Music; and so through Norah's sole agency, her elder son went from Eleven-plus failure, as the term was then, to classical pianist.

It is only now I look back in wonderment at what she did, this woman whose psyche had collapsed not all that long before; at the time, it was just what was happening around me as a teenager, and I took it for granted. Little by little, though, I could not but become aware myself of my mother's unusual qualities; as I started to learn about and appreciate the eighteenth-century Enlightenment, for instance, a period which she knew well, I began to realise that she herself epitomised the values which the Enlightenment had brought into being and which underpinned my own society. But it was her personal nature which won me to her completely, my dawning realisation of her gentleness and her generosity of spirit; she had a kindness which seemed without limit and was specially sensitive to the vulnerable, and eventually I saw what was informing it: at the age of twenty I had a very happy love affair which changed the way I looked at the world and when I hesitantly spoke of it to her I found she understood fully; you might say, as it seemed to me, that she understood about love.

From then on we grew ever closer, as I finished university and became a journalist, and she went beyond being a beloved mother, she was more, she was my best friend, and through the

1970s and into the 1980s I shared most things with her, from the characters I met in newspapers to the best of my generation's music: I played her much of Joni Mitchell, for example – *listen to this, it's really good* – and she loved it. She had a very open and adventurous mind. I thought she was incomparable. I was infinitely proud of her. I felt privileged, that she was my mother, she was the best thing I had – to be in a small suburban house with such intellect, such moral seriousness so lightly worn! Eventually, we began to talk of what had happened in 1954, which until then had been but a blur to me, and she told me about it as well as she could (although she could not speak of the core of it – that was something I had to discover for myself from her hospital notes) and I began to understand more about myself and especially, more about John. In 1977 he came home from four years as a pianist at the National Ballet of Canada and joined the Royal Ballet School, and when my father Jack, who was by then retired but no less irritable for that, began to pick on him, I defended him, for the first time; I began to try to be a proper brother. John had many difficulties by then; besides the emotional instability he carried from his childhood, he was an incipient alcoholic, and he was also gay, and still a practising Catholic, and the fact that the church appeared to condemn him for his sexuality as inherently sinful caused him agony. It was all an explosive mix. A particular crisis point came in 1982 after the dramatisation of Evelyn Waugh's *Brideshead Revisited* had earlier been shown on television. John particularly identified with the gay and Catholic Sebastian Flyte, and when he tried to discuss it obliquely with Norah, wholly innocent in matters sexual, she said that she didn't think there was *anything sinister* about Sebastian, and John went on the bender of all benders, which lasted more than a week, with him intermittently shouting in his drunkenness to her about *your sinister son*, to her great and bewildered distress, until I realised I had to intervene, and I sat her down and said to her, listen, there is something you have to know: John is homosexual. It was a shock to her. Her mind was formed long before sexual liberation and she barely knew what homosexuality was. It was certainly not something sanctioned by her religion, which she maintained to the end of her life. I gave her a fortnight, and then I went back to her and said: you have to accept it now. And she did. And she and John were able to talk about it, which was wonderful for him, and only just in time, for at the end of that year she was gone.

She was sixty-eight but her health had given out; a great physical burden was a broken vertebra, incurred when she fell out of bed – at the time, I was away in the Amazon. She had never been robust, but she grew frailer thereafter until she seemed like a tiny bird, and in the week before Christmas 1982 she had a stroke in her sleep. My father phoned me at seven in the morning, panicking as was his wont, crying, Michael, Michael, I can't wake your mother! She was alive, though deeply unconscious, and when I had helped my father do what needed to be done I drove across London to pick up John and we set off for Bebington, two hundred miles away up the M1 and when we got to the Watford Gap services area I knew that she had died, for I suddenly saw her in my mind bidding farewell, infinitely sad, and I phoned Mary at once and she said to me: 'God's taken her, Michael.' We were numb with shock and grief. All of us: me, John, my father, Mary and Gordon. Mary especially was deeply distressed, but she was formidably strong and began to organise the funeral. The following day I sat down and wrote an obituary notice which I put in the Deaths column of the *Liverpool Echo*, and which I have kept as a record of my feelings at that moment:

McCarthy – Norah: December 21, 1982, peacefully and fortified by the rites of the Church, in Clatterbridge Hospital, after a stroke, suffered at home, Norah McCarthy (née Day), beloved wife of Jack and adored mother of John and Michael and beloved sister of Mary and Gordon. Requiem Mass at St John's, New Ferry, Wirral, December 29 at 9.15 am. (She had grown increasingly frail and was in constant and severe pain from a back injury but her spirit was undimmed. Her perception was still acute and her common sense was still robust. Her intellect was still adventurous and her belief in civilised values was unshaken. She still had a stubborn optimism and an undying willing-ness to see the good in people. She was still completely free from pettiness or cant or falsity of any kind. Her gentleness and tenderness were as natural when she died as they must have been when she was a girl. The love she bore, for you, for me, for all of us, was the way Eliot described it, a condition of complete simplicity, costing not less than everything.)

I keep it as a record of my feelings, for then a weird thing happened: my feelings disappeared.

I suggested at the beginning of this book that we have expec-tations of our experiences, seeking to match them against para-digms; but experience does not always run in straight lines. Nor did it here. At first I thought it was just a mood, although a bizarre and disquieting one: I stopped feeling sad. My grief was replaced by indifference and complete unconcern that she was dead. My mother whom I adored, before all others: how could this be? To try to recover my grief, a couple of days later – it was Christmas night – I went to see her body, lying in its coffin, for I was sure that would do the trick, that would make me sad again. But the visit was even more disturbing: she was not in the body, she was not *there*, it was just a *thing*, and the sight of it triggered some savagely upsetting emotion deep inside me which I did not recognise; all I knew was, it shocked me greatly and drove my feelings further away than ever.

I did not have any frame of reference to account for this. Grieving is prominent among our customs, but having your grief suddenly and unaccountably disappear was not socially sanctioned by anything I was aware of, and I was bewildered, and upset, at not being upset: with my intelligence I knew full well what Norah merited, in terms of a tribute of love, I had written it in her obituary, but now I could feel it no longer in my heart. Varus, Varus, give me back my feelings! They did not come. It was the start of a strange period in my life, a lost period, I suppose, which lasted nearly a decade; for not only did the feelings for my mother disappear, but I soon found, in a way which was utterly unexpected and wholly inexplicable, that my sense of my own self-worth had vanished along with them. There was a complete collapse of self-belief, and at thirty-five, after what had seemed to be a pretty successful and confident existence, I was suddenly left with a sense of worthlessness, of being a moral vacuum entirely. I continued to function but I felt I no longer cared for any of the values which, for example, Norah had enshrined; I felt I no longer cared for anything, I felt I was merely standing in the ruins of my identity in a very unhappy way, and I struggled to understand it, year after year, but could not. It was just a fog, and eventually, seven years after her death, I grasped that if I were ever to find my way through it then I would need professional help. And so I began to talk about it all, climbing the stairs two evenings a week to the attic flat in Crouch End with the picture of Freud on the bathroom wall.

I will forever be grateful to the man who finally brought me the understanding and showed me the way out of the fog, but it took a long time: nearly three years more. It seemed to be an almost unbearably slow, painstaking process of unfolding layer after layer of emotions and of memories. But I possessed a signal advantage in John, as he had absolutely crystal clear recall of much of what had happened in the time of turmoil, which to me then was just a jumbled and confused mass, and I found that I could feed my slowly growing understanding with the detail he could bring back, distressing though it was to him; as the man I was talking to remarked, you've got the computer, he's got the data. The climactic moment of the whole process came in Wales, in the small coastal town of New Quay. I was on an assignment for *The Times* to go with Greenpeace to look at the dolphins of Cardigan Bay, and bad weather delayed the trip and I found I had a day to myself; I sat down in the hotel room and over several hours of intense thought I put everything together, everything I had learned, and in the evening finally discovered it.

I hated her.

I could hardly believe it. The mother whom I adored.

But there it was.

The hate was still there, at the deepest level of my whole being, in a place that had taken nearly three years to dig down to; I hated her for leaving me, in 1954, for saying nothing to me before she went, for not reassuring me or comforting me, but just abandoning me, just going. Yet the hate was something my psyche did not allow me to admit, and so, when I was seven, it morphed into indifference at her leaving; and in the same way, when she left for ever, when I was thirty-five, the mechanism kicked in again, and the indifference returned.

I hated her for abandoning me once more.

I hated her for being dead.

It was a great shock, but also a great illumination; and as soon as I got back home I went to the attic flat and poured it out, with the question I was desperate to ask: will it go now? The hate? Will it go away? Will it?

Andrea said, in his calm and quiet manner: 'I think, something changes.'

'What? What changes?'

'Perhaps,' he said, 'once you know what it is, it doesn't control you any more.'

Certainly, I felt, that . . . things were shifting. The log-jam was moving. But there was still a part of it all which troubled me, which was now out in the open, and the next week I went for dinner with John, and I said to him: 'You know, it really sort of bugs me that when Mum first went away to hospital, she didn't say anything to us about it. I mean, before she went. Like, you know. To comfort us. Or whatever.'

John said: 'But she did.'

'When? When did she?'

'She came up to our bedroom. She said, "I have to go away for a rest." I was crying. I said, "Don't go away, Mummy, please don't go away," and she said, "I have to, son."'

I said: 'I'm sorry, I don't remember that. I just don't remember that at all. I have no memory of it whatsoever.'

John said: 'You were asleep.'

'What?'

'You were asleep. She didn't want to wake you.'

'I was *asleep*?'

'Yes. She didn't want to wake you up.'

My head started to swim.

I saw this night, this August night, an age ago, where a family fell to bits: a mother in her anguish, being parted, perhaps for ever, from her two small boys, the elder one in tears but the younger one sleeping, unseeing, unknowing . . . and three days later, as soon as I could, I drove north, I drove to Bebington and stood by her grave, and there, at long last, the log-jam suddenly burst, and I wept for her.

To have recovered my feelings for my mother after nearly ten years, and to have understood why I lost them, and in so doing to have come to understand fully what happened in childhood, even though it had long seemed so confused and so obscure, was, as might be imagined, of great significance in my life, and a source of joy, not least as the mended love for Norah seemed even more complete; all that was missing, I felt somehow, was a marker of it, although what that might be, I did not know. I merely felt the common human need for meaning-making, for signing with ceremony of some sort the great events of life – birth . . . marriage . . . death – and now, the recovery of love. In the absence of any such marker, I tried to look with more empathy on the others who had been caught in the upset, beginning with John, and after he came to terms with his alcoholism, which was severe, we were able to explore his own truly tormented feelings about Norah's mental crisis and its aftermath with an insightful and patient woman who, impossible as it once might have seemed, gradually gave him something approaching peace. With the others – with my father Jack, with Mary and with Gordon – I talked to them as much as I could about what had happened, and found a similar attitude in all three of them, which was a sort of burning regret, that they had not acted as they should have done with Norah – they felt that, even though they had not properly comprehended what was happening or what her ailment was, in some way nevertheless they had failed her. I think at the heart of it was a consciousness of her qualities, of the love she bore, of her goodness, if you like, which they all felt they had themselves fallen short of, my father most acutely. Jack had come to realise what his sins were – they had been sins of omission. He had been a talented writer of light verse, and one day, quite unexpectedly, he put a piece of paper in my hand which said:

> There is no way I can forget
> The things I failed to do.
> Piaf found nothing to regret
> But I have much to rue.

My heart opened to him at once and I loved him, then, until the day he died.

They died one after another, in fact, in the last three years of the old century, and we buried them with Norah; the four of them lie together, Norah's love, it seemed to me, enfolding them all. I marked the grave with a headstone, and that was a worthwhile meaning-making, although underneath I still longed for something more, some fitting way in which I could commemorate for myself how truly exceptional she was, but I had to wait another ten years, for another generation; a guiltless one this time. In 2009 Jo and I brought Flora and Seb to the Wirral to see the grave, as they were old enough now, at seventeen and twelve respectively, to learn about the grandparents they had never known. It was a Sunday at the beginning of April; a cold morning of pale sunshine filtering through high clouds with a wintry north wind, though even in the chill air the cemetery was a pleasing place, lines of dark green cypresses and mature hollies giving it an Italian feel. We found the grave and the children read what was written on the headstone, and we stood in silence thinking about it, and as we did so a dead leaf came tumbling through the air towards us on the wind and fell at our feet, right at the grave's very edge. And then, in the thin sunlight, it opened its wings: it was a peacock.

I was taken aback.

I was taken aback just as I was by the swallowtail in Rimini and the morpho in the Amazon and the monarch in the garden in Boston.

The butterfly at my mother's grave.

It was ragged and rough after its overwintering, but the splendour of its colours was discernible still, the maroon wings and the four eyespots with their amethyst cores . . . and at once it set something alight in me, as butterflies have always been able to do, it set alight the fiery trail that led right back through my life to the summer of 1954 and the original time of turmoil, to the small boy gazing up at the buddleia. All day I thought about it, all through the journey back to London, and all of the evening, and the next morning I went into the office – the newsroom of the *Independent* – and offered a suggestion for a series of summer features, which the paper was always on the lookout for. I would try to see, I said, all the British butterflies – all fifty-eight species – over a single summer, and we would invite readers to join in and see how many they could find themselves, with a prize for the best entry. It was a good idea, and Roger Alton, the peerless editor (gone too soon, alas), and Oliver Wright, the sharp and energetic news editor, readily agreed, and the following week we launched the series with a magnificent butterfly wallchart the Indy graphics department produced, so captivating that when we later announced we still had some left, more than a thousand schools emailed us asking for copies. We called the series The Great British Butterfly Hunt and I started on it at once, seeking the help of Butterfly Conservation (BC), that most estimable of charities, which the chief executive, Martin Warren, willingly agreed to provide; and so, a month later, I found myself on the top of Butser Hill in Hampshire, at 888 feet the highest point in the South Downs, with Dan Hoare, BC's south-east England man, looking for the Duke of Burgundy. His Grace the Duke. The only British member of a worldwide butterfly family, the Riodinidae, the metalmarks; a true rarity. At seven years old I had thrilled to the romance of its name in *The Observer's Book of Butterflies* – no one knows how the butterfly came by it – but the wee beastie itself had always escaped me, and now it was very scarce and getting scarcer, and Dan had taken me to see it at one of its few breeding sites.

It was May Day. When I opened my eyes the bedroom window had shone brilliant blue and I was elated, but as the downs began to loom in the car windscreen on my approach I saw to my dismay that freezing sea fog had rolled in from the coast ten miles away and the hill-top was lost in misty swirls like a Lake District peak. Nonetheless, we proceeded to the summit. The fog slowly thinned, and on the very stroke of noon the sun burst through and began to warm up the local insect fauna, the odd bee, the odd hoverfly, and after a couple of minutes Dan called me over, and there it was, on a cowslip leaf amidst the hawthorn scrub – a miniature animal so bright in the lattice-work of orange and black on its wings that I had the impression of a glossy, freshly minted postage stamp. I was enthralled, and I said to my mother

Look.

The Duke of Burgundy.

This is for you

By then I had already seen about a dozen species, the commoner sights of the springtime, led by the peacock's three close relatives which also overwinter and also look like dead leaves with their wings closed: the red admiral, the comma, and the small tortoiseshell. I had found the two other lovely markers of early spring, the brimstone and the orange tip, and the first of the blues, the holly blue, and the first of the browns, the speckled wood, and the three common whites, the large, the small, and the green-veined, coming upon most of them in Kew Gardens near my home. But as the commoner species were ticked off, expeditions began to be necessary for the rarer and more difficult ones, and bigger expeditions than to Butser Hill. The biggest was to Scotland, to find the chequered skipper, the prettiest of the skipper family which went extinct in England in the mid 1970s but which survives and still flourishes in the hills of coastal Argyllshire. Thither I went in the company of BC's Scotland man, Tom Prescott, and he took me to Glasdrum Wood, rising steeply from the shore of Loch Creran, in the stunning landscape north of Oban. We were lucky with the weather, for there was just enough warmth and sunshine to bring the butterflies out – with the temperature a couple of degrees lower, we might not have seen them at all, and it was a very long way to go for nothing – although conditions were also ideal for the great affliction of anybody promenading in the Highlands, the Scottish midge, and so Tom produced his principal midge defence, an Avon bath oil called Skin So Soft, which the locals swear by. Generous applications of it became necessary as we climbed up through the wood and into a long open glade made for power lines, known as a wayleave, which is ideal habitat for the butterfly, as the midges swarmed about our faces, biting us like billy-o, but eventually we found it, a small brown creature exquisitely chequered with gold spots, and I said to my mother

Look at us.

Coated in Skin So Soft.

Bitten to bits by the midges.

But here's the chequered skipper.

It's for you

Argyll was the farthest trip, but it wasn't the highest: that was to find the mountain ringlet, our one true montane or Alpine butterfly, restricted to mountainsides in the Lake District and Scottish Highlands, usually at an altitude of between 1,500 and 2,500 feet, and this time I enlisted the help of two northern England representatives of BC, Dave Wainwright and Martin Wain, and the area they took me to was in the Lakes, starting from the Wrynose Pass which leads from Langdale over the mountains into Eskdale. We climbed steadily for an hour and a half, and we saw various butterflies, meadow browns or small heaths or painted ladies, but nothing resembling a mountain ringlet, even as the air got cooler and we began to glimpse the Lakeland peaks, with Skiddaw in the far distance; and the sweat ran down my brow and I began to feel we would be out of luck, until there was a sudden shout from both men, and I spotted a little whirring ball of black with an orange halo around it. It was a mountain ringlet, buzzing over the grass, and then

there was another, and another, and when we sat down to take
a breather, we found there was one resting in the grass stems
just behind us and we were able to look closely at its brown
wings, each with an orange band containing eyespots which
give the illusion of an orange glow when it flies, and eventually
I persuaded it to crawl on to my fingertip and we photographed
it there and I asked Dave Wainwright how high we were, and
the reading on his GPS was 614 metres, which is 2,014 feet,
and I said to my mother

See it on my fingertip?

At 2,014 feet up!

The mountain ringlet.

This is for you

To see every British butterfly species in a single summer was
a significant undertaking, I began to realise, and even though I
have met people who have done it unaided, in the strictly
limited time I had available I would not have managed without
Butterfly Conservation's expertise: for example, I had just half
a day set aside to see the swallowtail, and BC's Mandy Gluth
and Bernard Watts found it for me at How Hill Nature Reserve
in the Norfolk Broads, several swallowtails in fact, feeding on
the purple marsh thistles, so sensationally exquisite that I was
every bit as animated as when I caught sight of my first one
in Rimini forty years earlier, and I said to my mother

Look, look!

Swallowtails!

Nectaring on the marsh thistles!

Too beautiful to be true!

These are for you

There were many such pleasures, like the purple emperor,
and the Glanville fritillary and the white admiral and the silver-
spotted skipper and especially the large blue, which had
gone extinct in Britain in 1979 and which Professor Jeremy
Thomas, Britain's leading Lepidoptera scientist, had managed to

reintroduce, with great success; I saw it with Jeremy at Green Down in Somerset, murmuring

Look.

The large blue.

Back from the dead!

It's for you

But perhaps the most memorable moment of all was the dance of the heath fritillaries. *Melitaea athalia*, another take on orange and black, was one of our rarest species, but it had a stronghold in Blean Woods in Kent, a wide expanse of ancient forest near Canterbury and an RSPB reserve, part of which the warden, Michael Walter, managed specially for the butterfly. I had been there before and seen it with him, in moderate numbers, but that year his management – coppicing the woodland to get the right succession of vegetation to allow the larval food plant to flourish – paid off with spectacular results. Michael took me deep into the woodlands, about two miles down a track, then branched off along a narrow path which eventually opened out into a glade, a glade I thought you would never find by your- self in a million years: and there we came upon heath fritillaries in their hundreds, perhaps even in their thousands, fluttering everywhere a foot or so above the vegetation – they seemed to be dancing in the dappled light, thronging the sunbeams in a teeming minuet, in a silence broken only by birdsong, and I gasped at the sight and I said to my mother

See the dance?

It's a miracle.

The silent dance, in the secret glade.

This is for you

All summer I sought them out, week after week, and every sighting I wrote about in the *Independent* but first I dedicated to her; and then it was the end of August, and I had seen fifty- six species and only two remained. They were the brown hair- streak and the clouded yellow, and a delay in encountering the first was to be expected, as it is our last butterfly of the year to emerge, but the clouded yellow was a problem. Sulphur-golden blotched with black, it is an annual immigrant from the Continent, not at all uncommon, and I had seen many, but this summer I simply could not find it, and as August wore on I made half a dozen special trips into Surrey, Sussex, and Dorset without any luck; more than once I was told, you should have been here yesterday, and at length 31 August arrived, August Bank Holiday Monday, the final day of the summer and of The Great British Butterfly Hunt: the day of our reader's prize, which was to be a Butterfly Conservation-led safari to find the brown hairstreak (plus lunch).

The winner, out of many readers who had taken part, was an amiable, newly retired biology teacher from Friern Barnet in north London, Andy King, who had chalked up fifty-seven species, documenting and photographing them all, including the brown hairstreak already – a remarkable achievement – but who could not complete the set, as although he too had gone to Argyll for the chequered skipper, it had eluded him. Martin Warren and I took him to Steyning in Sussex where we met Neil Hulme of BC's Sussex branch, and in hazy warm sunshine Neil led us up into the downland to a strip of ash wood fringed with an understorey of blackthorn, where almost immediately we found a female brown hairstreak laying her eggs on the blackthorn stems, and I was stunned at the loveliness of this new species for me, chocolate-brown with broad orange-golden bands across her forewings, and I drank deeply of her beauty; and I made my dedication.

It was nearly noon. There was half a day left and I was still a clouded yellow short of the set, so we walked further up into the downs, the chalk grassland bright with pink wild marjoram and blue field scabious flowers, and over the next half-hour we saw various whites and small heaths and painted ladies especially – it was the year of a great painted lady invasion from the

Continent, and millions had arrived – and I was beginning to think it was simply one of those things and I would not do it when I heard a commotion and I turned to see Martin Warren waving his arms and he was shouting: 'Mike! Mike! Mike!'

'What?'

'Look!'

Bowling through the air was a flame, an intense sulphur flame, and there it was, the fifty-eighth and final species, on the last day of my butterfly summer, and I cried to my mother

See? See?

There it is!

The clouded yellow!

The last one!

It's for you

And somewhere inside me Norah laughed and said: *Yes, son, there it is!*

It was my gift to her.

It was my commemoration of who she was and how special she was; it was my tribute to the feelings which had been lost and found again, my tribute to their mended completeness; it was my final coming to peace. I gave her something which at last seemed appropriate to mark my love for her, my love which had gone astray in the time of turmoil all those years before, and so strangely, swerved off into butterflies.

I gave her all the butterflies of my country.

I gave her every one.

That the natural world can bring us peace; that the natural world can give us joy: these are the confirmations of what many people may instinctively feel but have not been able to articulate; that nature is not an extra, a luxury, but on the contrary is indispensable, part of our essence. And now that knowledge needs to be brought to nature's defence.

As we plough forward into the twenty-first century, like a ship heading into a gale, the threat which hangs over the natural world is without precedent: it points to a culminating moment in the history of humankind, this one life-form out of the earth's millions which, it may well be possible to write soon enough, rose from all the others to possess language and consciousness, to create art and law and medicine, even to voyage into space, but which ended up destroying its own home. No, we will not split the planet apart: the rocks and the seas will remain. But the natural life of the biosphere is now well on course to be devastated, in the Sixth Great Extinction of which we are the authors, just as it was in the previous five. Loss of habitat will see off countless species, as the tidal flats of the Yellow Sea, for example, despite supporting 50 million migrating shorebirds, disappear under concrete, as the rainforests continue to tumble under the chainsaws, as what's left of the mangrove swamps makes way for shrimp farms, and as all the remaining big rivers are dammed for hydroelectricity and their ecosystems perman-ently disrupted, while pollution on an ever-increasing scale, especially in the developing world, will be almost as potent a force for destruction as habitat loss. Over-exploitation of resources, over-hunting, and poaching will take a more and more critical toll of fish stocks and of larger animals such as elephants, rhinos, and tigers (a report from an international group of scientists in 2013 revealed that the African forest elephant, considered since 2010 as a species on its own, had lost two-thirds of its numbers to ivory poachers *just in the previous decade* and was firmly set on the path to extinction); and the menace of invasive species, fuelled by the ever-growing globalisation of world trade, will continue to wreak unexpected havoc (the Asian brown tree snake, accidentally introduced to the Pacific island of Guam after the Second World War, has wiped out nearly two-thirds of the island's native birds). Over everything hangs the spectre of global climate change, the greatest menace of all, now threatening to wreck the very stability of the atmosphere which, for hundreds of millions of years, has allowed life to arise and to flourish.

It has been difficult to be clear-sighted about the essence of what faces us, because the liberal secular humanism which has been our dominant creed for two generations, admirable though it is, has shied away from looking squarely at the issue: the fundamental, mushrooming clash between the earth and its species man, *Homo sapiens* − its problem child. In our current belief system, man is good, and so, as they used to say of General Motors and America, what's good for man is necessarily good for the planet; except, of course, that it isn't, and it may destroy the planet. The reluctance to see things as they are, while having to acknowledge nonetheless that the natural world is under assault − it can hardly be denied − invites a sort of displacement behaviour: it is tempting to locate the problem in particular political systems. But capitalism and the command economy share the blame for the despoliation of nature over the last century, and the natural world can be done down equally both by those after fast bucks for themselves or for shareholders and by those seeking the broader benefit of the commonwealth. Nature is not being destroyed just by a particular political or economic creed; it is being destroyed by the runaway scale of the human enterprise.

This becomes clearer if we look at a very basic concern for the future, how to feed the 9 billion people expected to be inhabiting the earth by the middle of the century. In 2011 the British government's Office for Science published a piece of horizon-scanning which addressed the issue head-on, entitled *The Future of Food and Farming*: it was a comprehensive look at the workings, now and in decades to come, of the global food system and how it could meet the challenges it will face. The report made several major recommendations: for example, that new technology which is controversial, such as the use of genetically modified organisms, should not be ruled out; and that the elimination of food waste should be of high priority. But one recommendation in particular caught my eye: the report proposed that no significant amounts of new land, such as tracts of rainforest, should be brought into cultivation, as this would release climate-changing greenhouse gases on a dangerous scale – in which case, 'the global food supply must be increased through sustainable intensification'.

After nearly seventy years of intensive farming, with everything that has been done to nature, in other words, the land that is in use for growing crops across the world will have to be made to work even harder still. The report showed itself aware of the environmental risks and the qualification 'sustainable' is important; but in essence, 'intensification' just means more and more chemical inputs, more fertilisers and especially more poisons, more 'cides – more pesticides, more herbicides, more fungicides, more molluscicides – and I found myself posing the sort of bizarre question which perhaps has never been framed in any human mind before: what does the twenty-first century hold for insects?

It seems to me that one of the prices of feeding 9 billion people in the years to come will be to sacrifice them. We may adore the charismatic megafauna, the snow leopards and the mountain gorillas, but very few of us are bothered about creepy-crawlies (other than butterflies and moths), which is doubtless why there has been so little awareness of the staggering decline in insect numbers that has emerged, in recent years, as a disturbing environmental phenomenon, indeed, as one of the defining ecological features of our age. Yet creepy-crawlies don't only creep and crawl; these are 'the little things that run the world', playing key roles in a myriad ecosystems, and their disappearance has profound dangers – finally recognised, of course, in the recent concern over the widespread vanishing of honeybees and other pollinators (two-thirds of our crops and fruit are wind-pollinated, but the rest need insect pollination). You may say, at least we will always preserve the pollinators; but I will wager you a pound to a pinch of snuff that there is a scientist somewhere, right now, probably in a pesticide company, toying with the idea of whether we can genetically modify insect-pollinated crops to make them able to be pollinated by the wind. No, insects will be surplus to requirements; they will have to go, as so much other life will have to go, so many species, so many habitats, so much of nature which has given us joy, as humankind appropriates to itself every bit of the natural world which it can grab; and the unmistakable signs are already there, with the insects especially, whose abundance in my own country has vanished in my lifetime, along with that curious phenomenon which once made their profusion so spectacularly manifest – the moth snowstorm.

What are we to do? We must feed our brothers and sisters. Who could argue against the alleviation of hunger? Which of us can so far step outside our species as to deny another person the right to eat? But what, then, about the earth, what if our needs as humans do indeed overwhelm it and consign its vibrant and wondrous life to the rubbish heap of history – what is our reaction to be? Too bad?

For nearly one hundred and fifty years now, people have been trying to defend the natural world in an organised manner: with their early appreciation of wilderness, which I have described, the Americans led the way, designating Yellowstone the first national park in the world in 1872. (The equivalent moment in Britain came considerably later, with the formation of the

Society for the Promotion of Nature Reserves for Britain and the Empire, by Charles Rothschild, in 1912, or perhaps Rothschild's earlier purchase of Wicken Fen in Cambridgeshire and his donation of part of it to the infant National Trust, with the stipulation that it be managed for the protection of the swallowtail butterfly.) Since then, the conservation movement has swelled to a substantial size, and become international: there is now a worldwide network of protected areas covering virtually every country, as well as a large group of wealthy, powerful, and committed non-governmental bodies across the globe that are dedicated to the defence of nature and its biodiversity.

It appears impressive. Perhaps even more impressive has been the role of individuals in caring for the natural world as the pressures on it began to mount. To give one striking instance from my own country: the lady's slipper, the blowsy, gaudy, most spectacularly beautiful of Britain's native orchids, was believed to have been driven to extinction in the early twentieth century by the rapacity of orchid collectors, until in 1930 a flower was found in the wild in a remote location. For the next forty years this one plant, which could have been uprooted and put in a large pocket in a matter of seconds, was protected tenaciously by a small group of botanists, whose principal tool was secrecy; and from 1970 its protection was overseen by a small government-backed committee (also secret) which organised round-the-clock volunteer wardening in the flowering season; until finally in the 1990s the Royal Botanic Gardens, Kew, worked out the difficult and hitherto-unknown trick of how to propagate *Cypripedium calceolus* in the laboratory, and the English lady's slipper was saved.

It was saved by a handful of people, all fired by their love of nature. And there have been other such cases. The love is real and can work wonders. But even their achievement is put in the shade by the sacrifices individuals have made, and are daily making, to defend the natural world away from the relative safety of Europe. For example, between 1990 and 2010 nearly 150 rangers were killed trying to protect the Virunga National Park in the Democratic Republic of the Congo, home to the majestic mountain gorillas, which are among the world's rarest animals but are located also in the centre of Africa's most terrible war; while, on another continent, at least 57 activists were killed in Peru between 2002 and 2014 trying to prevent the illegal destruction of the rainforest.

For it is becoming clearer than ever that despite the imposing size of the conservation movement, as things stand neither its individuals nor its organisations nor its aid dollars look like holding back the tide of destruction that is going to rise throughout the twenty-first century. There may be protected areas in every country, but delineating a protected area on a map and making it work properly can be two very different things, especially in the developing nations, and there, the illegal and often violent invasion of national parks, whether to take out their timber, mine their gold, kill their animals, or hack their forest down for farmland, is becoming one of the most heartbreaking aspects of the mounting assault on the natural world. The rhinos of the Kruger National Park in South Africa, for instance, began to be poached nearly a decade ago with the rise of the belief in Asia that rhino horn was a potent cure in traditional medicine; the subsequent surge in the slaughter, in this supposedly protected area, almost beggars belief. In 2007, 13 animals died; in 2008 the figure was 83; in 2009 it was 122; in 2010 it was 333; in 2011 it was 448; in 2012 it was 668, while in 2013 it reached 1,004. The reason why conservation is failing around the world is quite evident: it is the scale of the assault, itself a direct reflection of the extent of human numbers. It is not just parts of the natural world and its wildlife which are now at risk: it is nature itself. The protection is piecemeal; the threat is systemic.

At the outset, I criticised the two formal defences which have been put forward in a systematic way to counter the threat and to check nature's ruination, the pursuit of sustainable development and the argument of the worth of ecosystem services. I believe both are noble undertakings which have made and will make major contributions to conservation, but, as I said at the start, I believe they are flawed: sustainable development for depending on the goodwill of humans who may not themselves be good, and ecosystem services for being limited in what the concept may seek to protect. But it is time now to say that something even more vital is missing from them both: a belief.

Both can engage the intellect; neither can engage the imagination. In the late 1990s the British government gave £41 million to establish a high-profile world centre for sustainable development, the Earth Centre, built on a disused colliery site near Doncaster in South Yorkshire and intended to be a prestigious mass visitor attraction. There were hardly any visitors. It opened in May 2001, closed just over three years later, and is now forgotten. No one is going to stir the soul with sustainable development, no one is going to write poems about it, any more than they are going to write poems about TEEB (The Economics of Ecosystems and Biodiversity). These may be vitally necessary, but both are mere intellectual constructs; they can fill the minds of policymakers, but they cannot reach the hearts of people.

It is beliefs which do that; I mean, beliefs on the grand scale, faiths, and the briefest glance at history shows what beliefs which fire people's hearts can do. Consider the spread of Christianity, the spread of Islam, the power of the Renaissance, the power of the Reformation, the power of the socialist project. These are great events, but they are fully matched in historical significance by the calamitous event we are entering upon, the destruction of the natural world; and it seems to me, only a similar belief on the grand scale can hope to hold that back.

That belief, that faith, is available: it is the belief in nature's worth. People *are* going to write poems about that; they have done so for thousands of years. But something until now has been absent from the love of nature, from the delight in spring flowers and birdsong and the sense of the reawakening year, from the wonder at dolphins and the wonder at the dawn chorus: the modern understanding we are reaching, that there is an ancient bond with the natural world surviving deep within us, which makes it not a luxury, not an optional extra, not even just an enchantment, but part of our essence – the natural home for our psyches where we can find not only joy but also peace, and to destroy which, is to destroy a fundamental part of ourselves. Should we lose it, we would be less than whole. We would be less than we have evolved to be. We would find true peace impossible.

If we add that understanding to the love so many people already feel for nature, then we have, as it might be said, a new kind of love. It will be a love which is informed, but it will also be a love which, recognising the scale of the threat, is engaged, a love which, in delighting in a flower or a bird, or a meadow or a marsh, or a lake, or a forest, or a range of grasslands, realises it may not be there next year, and will do whatever it can to protect or save it; a love which can be fierce.

Such a belief, on the grand scale, can do great things, as even a single love like that has real worth; but thousands of loves like that have real power, since ordinary people's feelings are the beginnings of political will.

Now as the twenty-first century crashes upon the natural world like a tsunami, with all its obliteration and merciless unthinking ruin, let this new love be expressed; let it be articulated; let it be proclaimed.

Acknowledgements

Many of the themes in this book regarding the natural world and the grim threats now facing it are ones which I have discussed at length over recent years with a small group of naturalist and writer friends: Mark Avery, Tim Birkhead, Andy Clements, Mark Cocker, Peter Marren and Jeremy Mynott; and which I have discussed further in the alliance of the arts for the natural world, New Networks for Nature, set up by Messrs Birkhead, Cocker and Mynott, with John Fanshawe, in 2009 (and indeed the immediate spark for this book was a presentation I gave at the second N3 meeting in 2010 which dealt with the vanishing of abundance, entitled 'The Loss of Nature and The Nature of Loss'). I would like to thank them all, and many other members of N3 who share the same concerns, especially Katrina Porteous and Ruth Padel.

The list of other people who have helped me with *The Moth Snowstorm* and given generously of their time is, alas, too long for them to be thanked in detail so I can only name them. They are: Nick Askew, Phil Atkinson, Chris Baines, Helen Baker, Joanna Bromley, Mark Carwardine, Brian Clarke, Darryl Clifton-Dey, Franck Courchamp, Mike Crosby, Sarah Dawkins, Paul Donald, Richard Fox, Rob Fuller, Bob Gibbons, Lynne Greenstreet, Chris Hewson, Les Hill, Andrew Hoodless, Nigel Jarrett, Paul Knight, Georgina Mace, Graham Madge, Louise Marsh, Harriet Mead, Peter Melchett, Richard Moyse, Ian Newton, David Norman, John and Jane Paige, Debbie Pain,

Mark Parsons, Fiona Reynolds, Fiona Roberts, Chris Smith, Richard Smith, Denis Summers-Smith, Paul Stancliffe, Mike Toms, Paul Toynton, Gill Turner, Kate Vincent, Kevin Walker, Martin Warren, Cass Wedd, Colin Wells, Ian Woiwod and others.

Separately, I would like thank Nial Moores in Korea, as well as Spike Millington and Charlie Moores, and in Seoul, David Butterworth and Kim Jinyoung, who were enormously helpful; and I would like to thank the many people who helped me to see all fifty-eight British butterfly species in a single summer: Mark and Rosemary Avery, Robin Curtis, Clive Farrell, Polly Freeman, Mandy Gluth, Liz Goodyear, Dan Hoare, Neil Hulme, David Lambert, Andrew Middleton, Matthew Oates, Steve Peach, Tom Prescott, Jeremy Thomas, Martin Wain, Dave Wainwright, Michael Walter, Martin Warren, Bernard Watts, Ken Willmot and others.

For help with research and editing, I would like to thank Rebecca Lawrence and Marigold Atkey, as well as Simon Blundell, Librarian of the Reform Club, and Lynda Brooks, Librarian of the Linnean Society of London; and I am especially grateful to Ian Newton and Jeremy Mynott, who read the book in typescript. Any errors which remain are of course mine rather than theirs.

I must also thank Andrew Gordon and Roland Philipps, both of whom immediately saw the point, and made this book possible; and finally, as for what concerns myself, my thanks to Andrea Sabbadini and to Radhe Bentley, who effected the repairs, and to Jo Revill, who gave me the new beginning thereafter, are everlasting.

1982), reproduced by permission of Carcanet Press Limited; extract from 'A New Song' by Seamus Heaney from *New Selected Poems 1966–1987* (2002), published by Faber and Faber Ltd; extract from 'Coming' by Philip Larkin from *Collected Poems* (2003), published by Faber and Faber Ltd; extracts from 'The Hare' and 'All That's Past' by Walter de la Mare from *The Complete Poems of Walter de la Mare* (1971), reproduced by permission of The Literary Trustees of Walter de la Mare and The Society of Authors as their representative; extract from 'Prologue' by Dylan Thomas, from *The Collected Poems of Dylan Thomas: The New Centenary Edition* (2014), published by Orion, reproduced by permission of David Higham Associates.

Index